Content Preparation Guidelines for the Web and Information Appliances

Cross-Cultural Comparisons

Huafei Liao
Yinni Guo
April Savoy
Gavriel Salvendy

CRC Press
Taylor & Francis Group
Boca Raton London New York

CRC Press is an imprint of the
Taylor & Francis Group, an **informa** business

Human Factors and Ergonomics

Series Editor

Gavriel Salvendy

Published Titles

Conceptual Foundations of Human Factors Measurement, *D. Meister*

Content Preparation Guidelines for the Web and Information Appliances: Cross-Cultural Comparisons, *H. Liao, Y. Guo, A. Savoy, and G. Salvendy*

Designing for Accessibility: A Business Guide to Countering Design Exclusion, *S. Keates*

Handbook of Cognitive Task Design, *E. Hollnagel*

Handbook of Digital Human Modeling: Research for Applied Ergonomics and Human Factors Engineering, *V. G. Duffy*

Handbook of Human Factors and Ergonomics in Health Care and Patient Safety, *P. Carayon*

Handbook of Human Factors in Web Design, *R. Proctor and K. Vu*

Handbook of Standards and Guidelines in Ergonomics and Human Factors, *W. Karwowski*

Handbook of Virtual Environments: Design, Implementation, and Applications, *K. Stanney*

Handbook of Warnings, *M. Wogalter*

Human-Computer Interaction: Designing for Diverse Users and Domains, *A. Sears and J. A. Jacko*

Human-Computer Interaction: Design Issues, Solutions, and Applications, *A. Sears and J. A. Jacko*

Human-Computer Interaction: Development Process, *A. Sears and J. A. Jacko*

Human-Computer Interaction: Fundamentals, *A. Sears and J. A. Jacko*

Human Factors in System Design, Development, and Testing, *D. Meister and T. Enderwick*

Introduction to Human Factors and Ergonomics for Engineers, *M. R. Lehto and J. R. Buck*

Macroergonomics: Theory, Methods and Applications, *H. Hendrick and B. Kleiner*

The Handbook of Data Mining, *N. Ye*

The Human-Computer Interaction Handbook: Fundamentals, Evolving Technologies, and Emerging Applications, Second Edition, *A. Sears and J. A. Jacko*

Theories and Practice in Interaction Design, *S. Bagnara and G. Crampton-Smith*

The Universal Access Handbook, *C. Stephanidis*

Usability and Internationalization of Information Technology, *N. Aykin*

User Interfaces for All: Concepts, Methods, and Tools, *C. Stephanidis*

Forthcoming Titles

Computer-Aided Anthropometry for Research and Design, *K. M. Robinette*

Foundations of Human-Computer and Human-Machine Systems, *G. Johannsen*

Handbook of Healthcare Delivery Systems, *Y. Yih*

Human Performance Modeling: Design for Applications in Human Factors and Ergonomics, *D. L. Fisher, R. Schweickert, and C. G. Drury*

Practical Speech User Interface Design, *James R. Lewis*

Smart Clothing: Technology and Applications, *G. Cho*

CRC Press
Taylor & Francis Group
6000 Broken Sound Parkway NW, Suite 300
Boca Raton, FL 33487-2742

Printed in the United States of America on acid-free paper
10 9 8 7 6 5 4 3 2 1

International Standard Book Number: 978-1-4200-6777-4 (Hardback)

Library of Congress Cataloging-in-Publication Data

Content preparation guidelines for the web and information appliances : cross-cultural comparisons / authors, Huafei Liao ... [et al.].
p. cm. -- (Human factors and ergonomics ; 29)
"A CRC title."
Includes bibliographical references and index.
ISBN 978-1-4200-6777-4 (alk. paper)
1. Web sites--Design. 2. User interfaces (Computer science)--Cross-cultural studies. 3. Information organization. I. Liao, Huafei. II. Title. III. Series.

TK5105.888.C6588 2009
006.7--dc22 2009013807

Visit the Taylor & Francis Web site at
http://www.taylorandfrancis.com

and the CRC Press Web site at
http://www.crcpress.com

Contents

List of Figures

List of Tables

Preface

In usability studies of the Web and information technology (IT) products, what needs to be provided and how best to present the information provided are the central parts that determine the effectiveness of usability products and services—which ones the user likes and can perform in the least amount of time, with the least number of errors, and with the greatest ease. Previous publications about usability have concentrated predominantly on how to present the *what*. This book wholly concentrates on what information should be presented on the Web and for information appliances for different cultures.

The stimulus for this book germinated from the first three authors of this book. They each did their PhD dissertations on different aspects of content preparation requirements for Web-based design for American and Chinese populations and IT appliance design for the Chinese population. This book provides comprehensive coverage of content preparation, presents a discussion of each of the studies separately, and concludes with an integrated discussion of the subject matter and the derivation of guidelines for designing content preparation for Web and IT appliances. The notion of content preparation is much broader than has been presented in this book, as content preparation is needed for any decision making. In fact, it is content preparation that answers the question of what information is needed for making any specific decision.

This book is aimed at usability professionals and Web and IT appliance designers to aid them in better designing and evaluating what information should be included on Web sites and in IT appliance design. It should also be of use to researchers in the subject area. In producing this book, we wish to thank and acknowledge the contributions to these studies of Bruce A. Craig, Vincent G. Duffy, Mark R. Lehto, Robert W. Proctor, and Yu Michael Zhu.

Huafei Liao

Yinni Guo

April Savoy

Gavriel Salvendy

About the Authors

Huafei Liao is a control systems engineer at Bechtel Power Corporation. His current research interests are human factors and ergonomics and human aspects of information technology.

Yinni Guo is a PhD candidate in the School of Industrial Engineering, Purdue University. Her research interests are human–computer interaction, usability, and axiomatic design.

April Savoy is a research associate at SA Technologies. Her current research interests are human–computer interaction, situation awareness, and usability.

Gavriel Salvendy is emeritus professor of industrial engineering at Purdue University and chair professor and head of the Department of Industrial Engineering at Tsinghua University, China. He is a member of the National Academy of Engineering, and author/editor of 32 books as well as the author or co-author of over 270 journal publications relating to human aspects of design and operations of products and services.

1 An Overview of Content Knowledge

The objective of this book is to provide a theoretical foundation and operational tools to effectively prepare the content that needs to be presented so that a user is able to make correct decisions regarding the purchase of goods and services and be able to obtain information about items of interest. Content preparation is an integral part of usability. It answers the question what information needs to be presented for effective decision making. Once content preparation has been established, the question "*how* to present *what*" can be answered. If too much information is provided, then two problems may occur: (1) it may take too long to make decisions, and/or (2) it may increase the probability of making a wrong decision. On the other hand, if not enough information is available to the decision-maker, inappropriate suboptimal or incorrect decisions may be made. Content preparation as presented in this book will provide guidelines as to what information is needed for decision making by American and Chinese populations.

Human performance is generally affected by the capabilities of individuals, the level of training and interest they possess in doing the performance, and the characteristics of the design associated with what they have performed. Let us discuss briefly each of these items and conclude with a roadmap for the book as presented in Figure 1.1.

1.1 INDIVIDUAL FACTORS

Typically individuals vary in the performance capabilities 1:2 for 95% of the population (Salvendy and Seymour 1973). For example, the top 2.5% of the population could run a mile in, say, 6 minutes; then, for the bottom 2.5%, it would take 12 minutes or longer. Of course, this implies using the same population—for example, college students. Although people's ability is determined by heredity, environmental factors can maximize achieving heredity potential. One environmental factor that contributes to this achievement is personal training. Another is personal selection; this provides some assurance that the individual would have the attributes that allow them to be trained for the skills and the knowledge required for effective task performance. Also, it is important that there is a good match index between what jobs and tasks individuals would like to do and the actual tasks and jobs they need to perform. We need to be cognizant that individual's capability to perform tasks and jobs changes from day to day. Thus in large-scale studies, Salvendy and Seymour (1973) indicated that when correlating 1 day's performance with the next day's performance that there is usually a correlation of around .8 indicating that about one-third of the

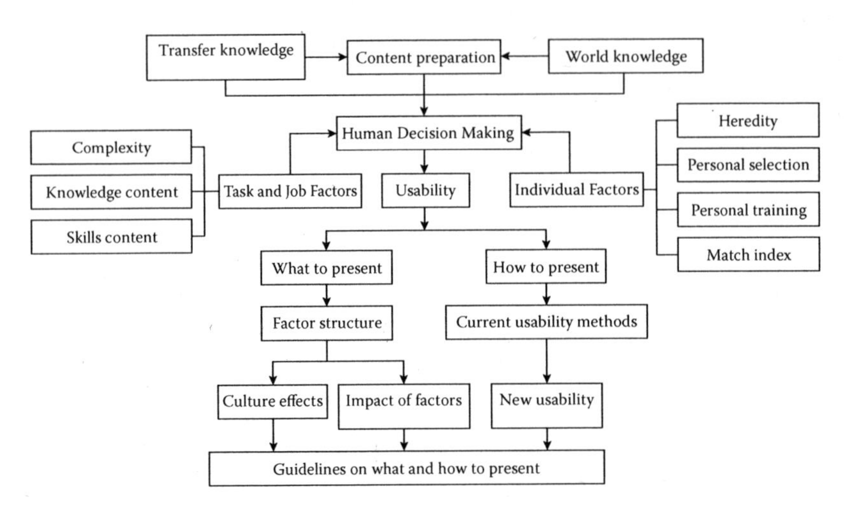

FIGURE 1.1 Derivation of guidelines for content preparation and an outline for the content of this book.

variables contributing to performance may be attributed to variation in the work setting and/or variation in individual's motivation and capability.

How individual factors influence what content knowledge is needed for the decision making is illustrated by individual differences within and between ethnic groups. This affects the type of information that is most important to provide to different user groups. The information needed for decision making may be obtained in a number of ways, including observation in a natural setting. Another approach for obtaining this information is using experimental studies with various scenarios to determine what information individuals use for decision making. A third approach is to obtain information using survey questionnaires, where individuals will indicate what they would like to use in making decisions. In this book, we have decided to use the third approach of questionnaires for the following reasons: It enables the collection of large sets of data for diversified populations and represents the individuals' overall experience as to what information is needed for decisions rather than the very targeted scenarios-based experiments, which provide information only on data presentation for very specific situations. The same would apply for observational studies. In dosing the survey studies, the data is usually analyzed by a factor analysis. Factor analysis is a statistical method that clusters items of similar attributes into a single cluster factor, where each factor refers to a different attribute. As indicated by Liu and Salvendy (2009), very careful attention needs to be given to internal consistency and reliability of the collected data from the questionnaires as input to factor analysis. Because data reliability significantly affects content preparation, we may incur both errors of omission and errors of commission regarding the nature of characteristics of information presented for decision making.

1.2 DESIGN OF TASKS AND JOBS

It is well known that the manner in which tasks and jobs are performed has an impact on their respective performance. For example, when menus are used, the performance time will depend on how many menu items are displayed, the width of each menu, and the distance the cursor has to be moved. This is based on Fitts' Law (Fitts 1954) and illustrated in Knight and Salvendy (1992), where it indicates that when the number of menus is the same, the distance of movement with the cursor is the same. However, the target width is twice as wide and that by itself causes a 10% decrease in overall performance time. Basically, the greater the number of alternatives from which a choice needs to be made, the longer it will take to make the choice. Also the greater the number of items presented, the greater the probability will be of selecting the wrong item or not making the best decision. Hence, it is important only to present content knowledge which needs to be considered for decision making and the operation of systems. To provide useful information in this regard, one must understand the complex nature of human decision making.

1.3 HUMAN DECISION MAKING

The quality of human decision making is affected by a variety of factors:

- Input variables have an effect on the number of variables; homogeneity and heterogeneity of the variables; chunked versus unchunked, continuous, intermittent, and intermittent or discrete variables; linearity and nonlinearity of the variables; subjective or objective derived measures for each of the variables.
- Impact of the outcome of the variables affects the decision process. If the outcome has a potentially large economic, personal, or other major impact, one tends to be more careful, take a longer time to make decisions, and tends to be more conservative than if the decision to be made has a low level of impact on the attributes considered.
- The process involved in how the decision is made also affects the outcome. For example, in the process of decision making, one may consider only the information presented relating to the specific case, where decisions need to be made. On the other hand, the decision maker also considers prior experience and knowledge on related cases and also considers world knowledge. In the latter case, the transferability of know-how from one scenario to another scenario is frequently subject to questions, and the issue is the potential value added for the effectiveness of decision making when information outside the domain associated with decision making is considered. The question is what information needs to be presented to people in order to maximize the effectiveness of the decision making.

1.4 CONTENT PREPARATION

Content preparation addresses the issue as to what information needs to be presented for effective decision making. It is important to present not too much or too little information since that would hinder effective decision-making. The information that needs to be presented is influenced by the mental model of the user. Since mental models of users differ across cultures (and to some extent within cultures), different information may need to be presented for individuals in different cultures. Once the decision is made regarding what information to present, then appropriate knowledge and methodologies have to be employed to present it in a way which results in easy, enjoyable and productive use.

2 Background Literature

2.1 INTRODUCTION

2.1.1 OVERVIEW

This chapter begins with a discussion about IQ and the quest to improve designs by highlighting different aspects of interfaces. Although human-computer interaction (HCI) introduced usability, the information systems (IS) research area developed design goals centered upon IQ. These designers had to adapt for power information users, where the focus of IS was the effectiveness of databases, search and retrieval algorithms, and content management systems. This different dimension of design brought information to the forefront and introduced terms of accuracy, timeliness, value added, and relevancy to system/interface evaluations. Hence, it provided a foundation for trends to develop in the area of user interface design and evaluation.

Furthermore, user-centered designs are often mentioned in the accompaniment of the term usability. Emerging into the mainstream in the 1980s, this term has acquired acclaim and research focus. Since then, it has been defined in many ways. The International Organization for Standardization (ISO) defines usability as the effectiveness, efficiency, and satisfaction of specified users with an interface in a particular context of use (ISO 9241-11). When people think of usability, the first things that come to mind are aesthetics and functionality. During the phases of design development, there is no doubt that attention and its resulting improvement to aesthetics and functionality are crucial. Consequentially, usability studies have delivered guidelines and evaluations methods that have been used to produce top designs.

Literature, as discussed in the following sections, has documented the evolution of user interface design aspects. This chapter discusses the need to consider information and usability in a design approach and presents content preparation as an example. This theory identifies information that is needed and how to organize, structure, and present that information so it can easily be retrieved when needed (Proctor et al. 2002). The development of content preparation stems from established success of usability and IQ in user interface design.

The best way to illustrate the strength of content preparation is to identify its application to, Web sites, and information appliances. E-commerce, advertisement, and education Web site studies illustrate the use of this theory in conjunction with Internet technology that has become a daily essential for millions of individuals. In addition, information appliances are used to display the vigor of content preparation. The following sections present existing studies of the aforementioned application domains to explain the need for a design approach that considers aesthetics, functionality, and information as equally important.

Lastly, the illustration of content preparation's impact on design does not end with the demonstration of application domains. This chapter concludes with consideration of the tremendous amount of diversity among users from a cross-cultural perspective. A majority of studies consider the traditional capabilities of humans (e.g., workload, anthropometrics, information processing) when discussing user demographics. With applications such as the Internet and the Web sites that it supports, the pool of users differ in regards to language, education, culture, etc. In that, varying cultures have proven to be a hot topic as technological advances decompose geographic interaction barriers. This issue becomes increasingly significant as more companies begin to market their products and/or services across national and cultural boundaries on their multilingual Web sites. Today, merely translating a Web site to another language is not enough in international e-commerce. People in different countries will often approach similar tasks in different ways because of cultural differences (e.g., Degen et al. 2005; Honold 1999), and thus it is important to take cultural and psychological differences between countries into consideration in international interface design.

2.1.2 Information Quality

Data and information quality (IQ) have been a concern and research focus across many disciplines (Katerattanakul and Siau 1999; Pipino et al. 2002; Strong et al. 1997). Each area has its own approach for classifying important components and evaluating quality. For instance, the area of IS centered its approach upon the effectiveness of databases, search and retrieval algorithms, and content management systems (Ives et al. 1983). On the other hand, HCI approaches are founded on usability and customer satisfaction (Baierova et al. 2003; Zhang et al. 1999). Despite the differences of approach and aims, there is one constant—the high likelihood that enhanced information would have a positive impact on designs (Savoy and Salvendy 2007). However, there is an insufficient amount of focus on content preparation. That, this section investigates studies in IS in hope of discovering a construct defining useful content for general use.

Studies in IS have focused on the development classification schema for information (i.e., data) quality assessment and benchmarking (Dedeke 2000; Delone and McLean 1992; Knight and Burn 2005). Overall, quality information is defined as information that would aid users in their tasks, decision making, and duties. There are specific attributes that help in the determination of quality information. Over the years, there has been a number of classification frameworks developed to allow IS managers to better understand and meet the needs of their information consumers (Ives et al. 1983; Strong et al. 1997). Quality category is one example. It consists of four classes of information: intrinsic IQ, representational IQ, accessibility IQ, and contextual IQ (Knight and Burn 2005).

The intrinsic IQ category represents the need for information to have an independent quality (Huang et al. 1999). This category encapsulates attributes such as accuracy, coverage, and currency. *Accuracy* is the extent to which data are correct, reliable, and certified to be error-free (Alexander and Tate 1999; Katerattanakul and Siau 1999; Strong et al. 1997). *Coverage* is the range of topics included in a work

(Alexander and Tate 1999). *Currency* is the extent to which material can be identified as up to date (Alexander and Tate 1999). These three attributes are commonly shared among IS classification theories for data quality. They present the basic necessities of general content IQ requirements.

Representational IQ addresses the format and presentation of information. The visual (e.g., how) aspect of interface design is essential to user satisfaction. Thus, this category is often represented across the research domains. It will not be discussed in detail here because it is not specific to what information should be presented only how to present information.

Security and privacy are the main focuses of accessibility IQ (Ives et al. 1983). This category includes attributes such as security, objectivity, and authority. *Security* is the extent to which access to information is restricted appropriately to maintain its security (Kahn et al. 2002; Naumann and Rolker 2000; Strong et al. 1997). Objectivity is the extent to which information is unbiased, unprejudiced, and impartial (Alexander and Tate 1999; Kahn et al. 2002; Naumann and Rolker 2000; Strong et al. 1997). *Authority* is the extent to which the material author/company has definitive knowledge of the given subject area (Alexander and Tate 1999). This category could be viewed as trust IQ.

Contextual IQ highlights the variation of consumers' need for information according to their tasks. This category comprises four general categories: relevancy, value-added, timeliness, and amount of data. *Value-added* is the extent to which information is beneficial, provides advantages from its uses. *Relevancy* is the extent to which information is applicable and helpful to the task at hand. *Completeness* is the extent to which information is not missing and is of sufficient breadth and depth for the task at hand. *Amount of information* is the extent to which the quantity of volume of available data is appropriate.

Table 2.1 displays the content IQ and trust IQ categories and their respective information content requirements. The information content requirements are specific pieces of information that should be included to achieve a particular type of IQ. With this level of detail, these requirements could be included in evaluations and measured easily for a wide range of interfaces. Currently, not many (i.e., less than 5%) usability and customer satisfaction evaluations include measures of information content. When information is considered in the evaluations, the measures are high-level or overall questions such as, Was the information was of high quality? How appealing was the subject matter? (Hornbaek 2006). Although this type of assessment is essential, more information is required to pinpoint the information problems and develop solutions.

Detailed content requirements for attaining general content IQ and trust IQ provide insight for defining/determining and assessing credible information. Designers should be mindful that credible information and useful information are not the same. Although useful information should be credible, credible information is not necessarily useful. Discussions of contextual IQ describe it best, where information that is relevant or value-adding may be credible and useful in one task but only credible or neither in another task. Although a list of content requirements may not be applicable to every interface, a list should be developed for the various, related interfaces. This warrants research to define the content requirements,

TABLE 2.1
Dimensions of Information Quality Pertaining to Content Usability

Category Type	IQ Dimensions	Content Requirement	Sources
Content IQ	Accuracy	Information source Editor name Process independent evaluations of products	Alexander and Tate 1999; Kahn et al. 2002; Katerattanakul et al. 1999; Klein 2002; Naumann and Rolker 2000; Strong et al. 1997
	Coverage	Detailed description for the products and services offered Intended audience	Alexander and Tate 1999
	Currency	Date of creation/revisions	Alexander and Tate, 1999
Trust IQ	Security	Security measures for financial transactions Downloading restrictions	Kahn et al. 2002; Naumann and Rolker, 2000; Strong et al. 1997
	Objectivity	Relationship between company and advertisements Site's policy relating to sponsorship/advertisements Goal of the organization/ mission statement Names of nonprofit or corporate sponsors	Alexander and Tate 1999; Kahn et al. 2002; Naumann and Rolker 2000; Strong et al. 1997
	Authority	Name of responsible organization Contact information for organization Qualifications of the organization Copyright holder information Ratings or recommendation Names and qualifications of company significant employees Presence beyond the Web Nature of the company/organization How long the company has been in existence Refund policy Warranty	Alexander and Tate 1999

Source: Data from Savoy, A and G. Salvendy, 2007. Effectiveness of conent preparation in information technology operations: Synopsis of a working paper. *Proceedings of the 2007 Human-Computer Interaction International (Part I)*: 624–31

which provide the same level of detail as the content IQ and trust IQ dimensions, for useful information—relevant, value-added, and completeness. From the findings, a clear definition of useful content cannot be deduced from the studies in this area alone.

2.1.3 Usability

The term usability has been in use since the early 1980s. Since its birth, it has acquired related terms such as user friendliness and ease of use (Hornbaek 2006). The establishment of a comprehensive, universally accepted definition for usability has been difficult. Lewis (2006) attributes the complexity to the issues that usability is not a property of a person or thing and there are two conceptions of usability. *Usability* is a property dependent on interactions among users, products, tasks, and environment. The International Organization for Standardization (1998) defines usability as the "extent to which a product can be used by specified users to achieve specified goals with effectiveness, efficiency, and satisfaction in a specified context of use" (p. 6). Comparatively, Chapanis (1981) defines usability as being "inversely proportional to the number and severity of difficulties people have in using software."

The development of dimensions was addressed in a recent review of current practices investigating how usability is measured. Hornbaek (2006) noted in this review that the meaning of the term usability varied greatly among researchers. The definition was based on whatever items could be measured. Over the years, many researchers have expressed definitions for usability, which were used as guidelines for measurement. Shackel (1991), Nielsen (1992), and ISO (1998) have developed definitions of usability that are among the most referenced (Ryu 2005). With these definitions, these authors established criteria or dimensions to eliminate ambiguity and aid in selecting measures. Figure 2.1 illustrates the comparison of usability dimensions of the definitions referenced by Ryu (2005).

2.1.3.1 ISO 9126

The ISO 9126 (1991) standard has a focused perspective of usability. It defines usability as a "set of attributes that bear on the effort needed for use, and on the individual assessment of such use, by a stated or implied set of users" (Bevan 1999, 1). This standard's objective is to meet the needs of the user with emphasis on quality, particularly software quality.

Software quality is decomposed into six categories of characteristics: functionalities, reliability, usability, effectiveness, maintainability, and portability. Bevan (1999) recommends the use of the revised version of this standard, ISO 9126-1, which includes a quality model identifying three approaches to product quality. These approaches are listed as internal quality, external quality, and quality in use. Of the three, quality in use has the most similarity with ISO 9241-11. It measures the extent to which software meets the needs of the user in the working environment. However, this model does not explicitly mention the information needs of users.

The definition of usability in ISO 9216 provides a starting point for identifying the set of information components needed for content preparation.

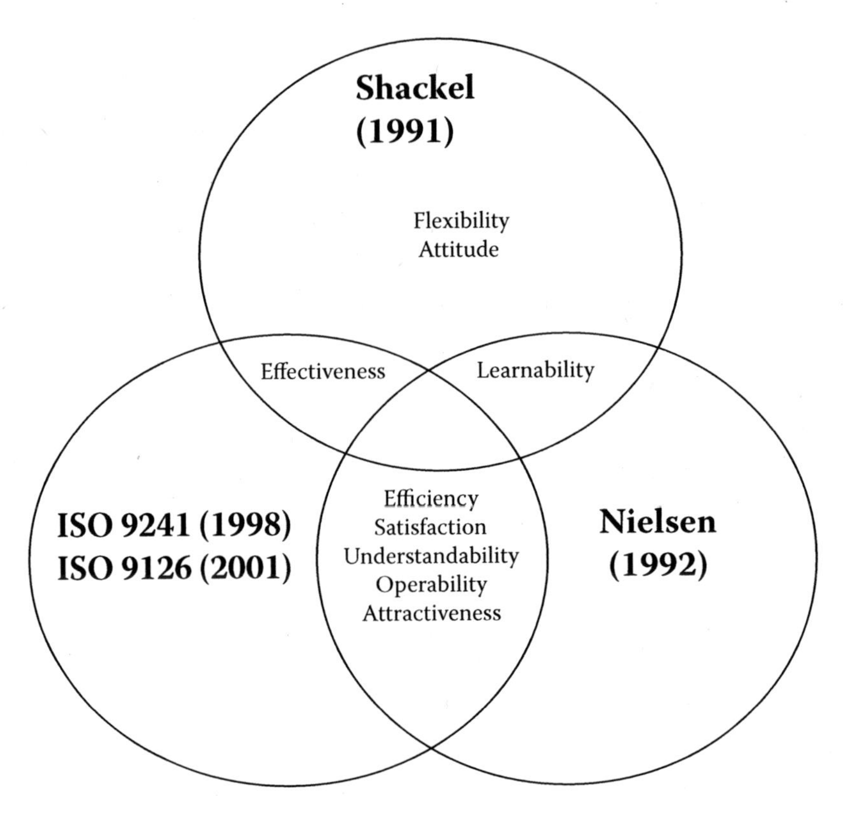

FIGURE 2.1 Comparisons of usability dimensions and definitions. (Modified from Ryu, Y. S. 2005. Development of usability questionnaires for electronic mobile products and decision making methods. PhD dissertation, Virginia Polytechnic Institute and State University.)

2.1.3.2 ISO 9241-11

The ISO 9241-11 standard is entitled *Ergonomic Requirements of Office Work with Visual Display Terminals: Guidance on Usability* and is one of 17 parts dedicated to ergonomic requirements pertaining to visual display terminals. The main focus of Part 11 is usability and its dependence on the rest of the work system/environment. It advocates a process-oriented approach to usability (Abram 2003), offering general guidelines and techniques.

The ISO 9241-11 standard has a broader view than ISO 9126 does. It does not specify requirements to use particular products or methods. Instead, it enables the measurement of user satisfaction and performance with which a product is usable in a particular context, which is characterized by the users, tasks, environment, and equipment. This provides the advantage of holistic evaluation of effectiveness, efficiency, and satisfaction explaining the aspects of usability and the relationships between them.

Guidance provided by this standard can be used to identify the users, tasks, and environments so that more accurate judgments can be made about the needs for particular product attributes as highlighted in ISO 9126. Product usability evaluation and design can often be improved by incorporating features and attributes known to benefit the users in a particular context of use. Consequentially, a product can have different levels of usability when used in different contexts. Therefore, it is necessary to measure the performance and satisfaction of a product in solidarity and while users are working with a product.

2.1.4 Definition of Content Preparation

Content preparation emphasizes information requirements, the identification of specific information needed by users and/or customers, during interface design. Its attention to information is its trademark. The conceptual structure implies that consideration of information requirements in the early stages of design will enhance interface aesthetics, functionality, and overall user experiences. It combines the user-centeredness of usability and the information emphasis of IQ to develop a theory that strives to provide the user with the information needed to complete a task and improve decision making.

Content preparation, which has been recognized as an important stage in Web site development, is defined as the process involved in determining the information for the Web site to convey, and how to organize, structure, and present that information so that it can be retrieved easily and efficiently when needed (Lazar and Sears 2006; Vu and Proctor 2006). For sites in certain domains, such as a library catalog, content preparation in terms of these four aspects is clear. This is so not only because the organization and structure of information, as well as the type of information to be displayed, is relatively straightforward (Proctor et al. 2002, 2003), but also because a vast amount of research effort has been put into these domains to study the issues of content preparation. However, for sites in other domains such as e-commerce, e-health, and e-education, this often is not the case due to the following issues (Proctor et al. 2002, 2003; Van Duyne et al. 2003):

- The content designer does not know what types of information need to be conveyed to the end-user and what techniques are appropriate to elicit it.
- The content of these Web sites cannot be readily classified into well-established categories and organized in a logical way for search.
- It is not obvious which methods are most effective and efficient for retrieving information in users' interactions with Web sites.
- The web designer does not know how to present the information in a manner compatible with users' goals.

Proctor et al. (2002, 2003) discussed content preparation in a broad sense and identified four aspects: knowledge elicitation, information organization and structure, information retrieval, and information presentation (see Figure 2.2). These aspects address the aforementioned issues regarding application of content preparation with existing and new web domains and other information technologies. Knowledge elicitation and information presentation pertain to the front end of interface design. Information organization and structure pertain to the back end of interface design. Information retrieval pertains to a particular functionality of interface design.

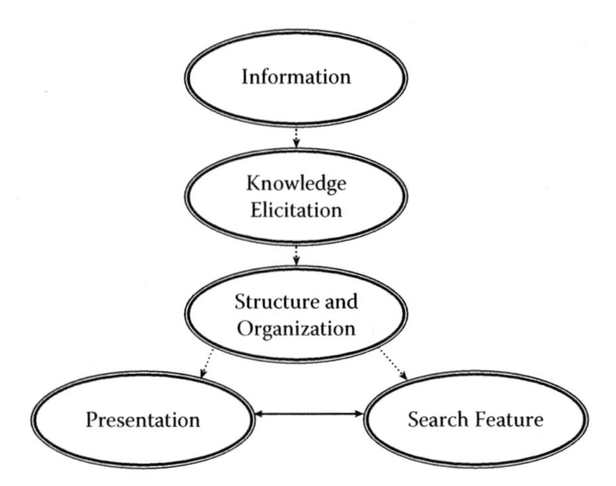

FIGURE 2.2 Schema for content preparation and management. (Modified from Proctor, R. W., K.-P. Vu, G. Salvendy, et al. 2002. Content preparation and management for Web design: Eliciting, structuring, searching, and displaying informaion. *International Journal of Human-Computer Interaction* 14(1): 25–92.)

There may be some confusion between knowledge elicitation and information presentation. Both components are vital to successful interface design; however, one component has received more research focus and acknowledgment than the other. Information presentation is associated with aesthetic design and has been very popular over the years. Knowledge elicitation is associated with the determination of what information should be presented. It is important to note the differences between the two. *Aesthetic design* concentrates on elements such as font size, color, position. *Knowledge elicitation* is concerned with what type of information is needed on the screen: for example, price, address, and warranty information. Content preparation front-end components focus primarily on two basic issues related to the information content of a Web site: (1) determining what information content is needed by users, and (2) determining how to present that information (Lazar and Sears 2006). The goal of aesthetics design is to make a Web site visually attractive and enjoyable. In contrast, the goal of knowledge elicitation is to provide the user with the information needed to complete the tasks associated with a particular interface, thus improving user information satisfaction.

How should content designers determine the information needed for interfaces? Knowledge elicitation is the component of this theory that addresses this question. It involves identification of knowledge structures and processes that competent individuals bring to task performance (Proctor et al. 2002). A designer must identify sources of information, and there are three popular sources—experts, users, and metadata. Each of the sources offers designing essentials. Experts and targeted users are important due to the variety of mental models and/or scripts pertaining to processes used in completing tasks; therefore, knowledge elicitation from both groups ensures a more comprehensive design (Proctor et al. 2002). Variety within each of the groups is also beneficial because mental models differ among different user groups and expert levels. Incorporating the information from the two sources will enable the developers to predict and satisfy expectations of the interface. Metadata is also a good option, where developers analyze data to gain insight about information requirements. Data could be used to identify trends and design gaps.

The back-end and functionality components of this design are organization and structure, and information retrieval. The organization and structure component of this design theory directs attention to the back end of system design. It reiterates the idea that a system should organize information in a clear and logical way, which has been suggested by information architecture. Information retrieval focuses on the search engine capabilities of an interface. This component has received increased attention with the development of new and popular search engines such as Google. Its design concerns include search algorithms and the presentation of their results. Even though these components are crucial in system overall design, the back-end and functionality aspects of content preparation will not be discussed in this text.

Information determination is the root of this theory and its advances influences all the other segments (Savoy and Salvendy 2008). In addition, information presentation provided a direct link to this theory and usability that illustrated its significance in the well-established research area. Determining what information to display on interfaces is a complex task. This decision must consider the design and user goal, the demographic of the users, the capability of the technology, and overall usability. Throughout this book, in-depth discussions and experimental analysis will be presented. Insights will be revealed regarding different user groups, elicitation methods, and information design guidelines for information technology.

2.2 CONTENT PREPARATION FOR THE WORLD WIDE WEB AND INFORMATION APPLIANCES

2.2.1 World Wide Web

The World Wide Web (WWW) has proven to be to the most popular information transfer medium of the twentieth century. Approximately 30 years after the invention of the Internet, WWW extended the use of information sharing technology from limited academicians and military to the rest of the world. The world's Internet usage and population statistics show interesting trends. Asia and Europe represent the highest percentage of use among the world regions. North America is third. However, the story is in the usage growth reported: from 2000 to 2008, usage growth in Africa and the Middle East was over 100% (Miniwatts Marketing Group 2008). This depicts the diversity of the WWW users and the large set of novice users as well.

With the push of buttons, immeasurable amounts of information can be accessed. E-environments have changed the way people shop, communicate, bank, and so on. Interfaces designs implementing usability guidelines can ease user transition to e-environments. Research will support that user satisfaction increases with interfaces that are easy to use and aesthetically pleasing. This research is highly concentrated with issues of presentation and functionality. What about user information satisfaction? Information plays an integral part in completing tasks. If the World Wide Web is an information sharing tool, the information that it presents should be evaluated. Simply presenting all of the information that can fit on a screen or is allowed by bandwidth does not produce a usable interface. The desired information is that which informs users and aids in the completion of their tasks. Anything else is classified as "noise."

Within each environment, certain information is needed by users. For example, e-banking users want to manage their finances; therefore the type of information that is important to them includes account balance, bill payment, interest rates. The decisions they would make include: Should I transfer money? Should I check my account activity? E-healthcare is different because users are trying to decide "Should I go to the doctor? Should I get medications, etc.?"

Content preparation emphasizes information that will aid users in their decisions and tasks. A number of studies are restricted to specific domains, which include financial, e-commerce, entertainment, education, medical, and government. The results of each study provide in-depth analysis of what aspects/features were important to usability and customer satisfaction. However, there is consistent lack of literature concerning information content. It is true that user preferences and information needs change according to their tasks and/or e-environments (Baierova et al. 2003; Zhang et al. 2001), yet, the desire for useful information is common for users across the varying types of e-environments.

2.2.1.1 E-Commerce

E-commerce is a highly competitive e-environment. With a well-designed web interface—information, functionality, and presentation—e-business has the potential to be lucrative. Amazon.com, Travelocity.com, and Ebay.com are a few examples of success stories in this e-environment. Conversely, many have fallen victim to stiff competition—the dot.com burst. Due to the overwhelming number of existing Web sites with the same identical target audiences, there is little room for error in the design of the user experience. Customer loyalty is not as concrete as it was in the shopping days of yesteryear. When a market is saturated by e-commerce Web sites offering similar products/services and targeting the same potential customers, the margin for error with regards to meeting (or surpassing) customer expectations is minute. The amount of customer loyalty has decreased over the years with the emergence of e-commerce. Previously, customer loyalty was determined by location, personal relationships, and traditions. E-commerce has dramatically decreased the need to travel, limited human interaction, and foregone traditions. The lack of loyalty has motivated customers to shop around more. Thus, companies are constantly looking for something that will give them the edge over their countless competitors and attract new and repeated business from e-consumers (Savoy and Salvendy 2008).

For e-commerce, information-centered Web site designs are vital to success. The information presented on these e-commerce Web sites is of extreme importance. It can positively (or negatively) influence e-consumers' purchase intent and shopping satisfaction (Ranganathan and Ganapathy 2002). Designs that incorporate useful information can increase sales, reduce returns, and motivate users to return for future purchases (Lazar and Sears 2006). Web sites lacking information-influenced designs have costly effects. On such sites users are unable to complete transactions without supporting information such as help aids, purchase information, shipping information, and error messages. Sales are lost due to lack of useful information and/or noise. Potential customers may leave if it takes too long for them to find the desired information. This reiterates that information must be filtered. Too much information causes increased search time and user frustration. Do not display information simply because there is white space on the webpage. Display information because it will assist users with their tasks. According to Nielsen et al. (2001), 36% of the time, users cannot find the information that they are searching for on a webpage. From this and other factors, search algorithms and functionality have received much attention. Although advancements in those areas will help the problem, advancements in content preparation and the development of a definitive framework for useful information would lead to design breakthroughs.

Although some studies have also offered design guidelines regarding the information content to include on e-commerce Web sites (e.g., Fang and Salvendy 2003; IBM 2004; Rohn 1998), information content—the what—has often been neglected as a design component. In recent years, proposals for effective e-commerce design have been published; however, their models focus primarily on information presentation (e.g., menu structure, page layout) (e.g., Tullis et al. 2005) and information architecture (e.g., Vu and Proctor 2006). E-commerce Web sites should serve as major sources of specific information regarding products and/or services, customer support, privacy policies, and other important issues (Aladwani and Palvia 2002; Ranganathan and Ganapathy 2002). Quality criteria could be applied in design because the information content on an e-commerce Web site should be relevant, understandable, reliable, adequate, useful, and should cover a broad scope (McKinney and Zahedi 2002). Designers should ensure that Web sites contain what consumers want and need (Loiacono 2000), and that online consumers can find the desired information easily so that they will not be likely to abandon their efforts (Lazar and Sears 2006). These general guidelines are beneficial, but they are limited because they only cover part of the information content on e-commerce Web sites. It is documented that customers need specific information that will aid in their decision making.

Are there detailed lists of information requirements, specifying what will aid users in their decision making, that designers can use during development? Some lists have been created that are tailored to e-commerce (e.g., Guo et al. forthcoming; Liao et al. 2008; Savoy and Salvendy 2008). For instance, e-consumers want the following information: available alternative brands, evaluation criteria for brand comparison, importance of evaluation criteria, and information on which to form beliefs (Mowen and Minor 1998). Detlor et al. (2003) examined consumers' information preferences for browsing and searching e-commerce Web sites. Always have the consumer in mind; consider their task (what) and their approach (how). Do your information needs vary when you are searching for a particular item versus when you are just browsing? Well, a three-level hierarchical structure was organized considering the two approaches (Figure 2.3). The top level provides a high-level description where the components are listed as product-related, retailer-related, and interface-related information (Detlor et al. 2003). The second and third levels of the structure provide increasingly more details with information cues and attributes, respectively. This structure displays the information intensity of e-commerce Web sites.

While consumers' primary needs in this e-environment are related to aspects regarding products/services, information related to the interface and retailer are also important. Other researchers (Barnes and Vigen 2001; Gehrke and Turban 1999) suggest that

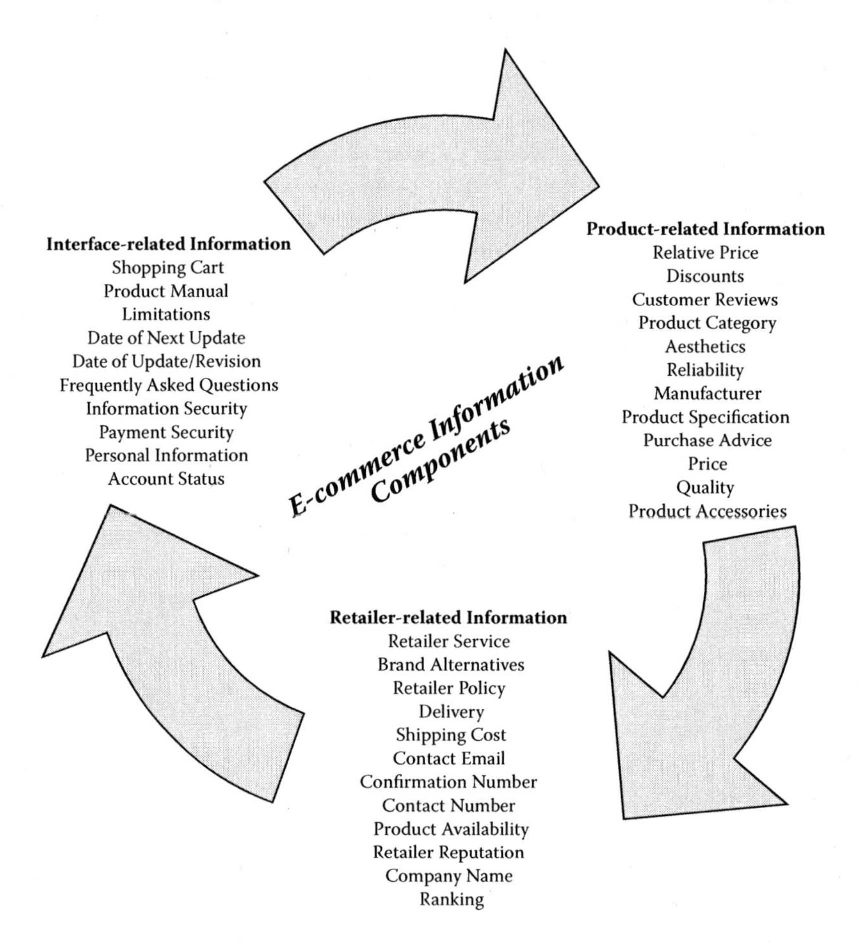

FIGURE 2.3 Essential information components in e-commerce domain. (Modified from Savoy, A., and G. Salvendy, 2008. Foundations of content preparation for the Web. *Theoretical Issues in Ergonomics Science* 9(6): 501–21.)

consumers want to know interface-related information such as account status, security issues, and payment alternatives. In addition, consumers need to know information about the retailers such as phone numbers, mailing address, and reputation.

A framework clearly identifying users' needs would provide designers with the tools needed to enhance customer experiences. How does this type of research translate into practice and successful designs? Researchers have confirmed that users value information regarding price, options, and comparisons. In fact, comparisons are high–priority information, which have a great impact on purchasing decisions. Constructing a Web site built upon these information components would lead designers to price-comparison Web sites. Design Web sites with a plethora of options and allow price comparisons. There is no doubt that this would work. It is the secret to the success for Orbitz.com, Expedia.com, and Dealsea.com. Furthermore, price does not have to be the only type of comparison. Web sites have compared options based on additional services, delivery options, and other factors. A Web site could offer customizable comparisons where the customers could choose the comparison attribute dynamically. For example, Addall.com shows all available books (the same book) from different e-business Web sites and offers some attributes for comparison.

2.2.1.2 E-Advertising

Online advertisements have been known to boost revenues by three billion dollars (Greer 2003). Advertisements are geared toward the promotion and sale of a product, service, idea, or company (Jones et al. 2005). Furthermore, the mechanics of advertising is heavily dependent on the information communicated (Mueller 1991). Popular design approaches—informative and emotional appeal—have been effective in traditional and nontraditional media, which demonstrates the vitality of the content as the media evolved. Advertising models and theories have been dedicated to explaining how a consumer searches for functional information to assist decision making during the purchasing process. Interestingly, the connection between advertisements and online shopping has transitioned from traditional advertisements and shopping dynamics. Due to the strong connection, there are redundant elements of information content between the two; however, the presentation differences (e.g., online banner ads) influence the information needs that are unique to e-commerce and e-advertisements (Savoy and Salvendy 2008).

Since the 1970s, research conducted in the marketing area addressed needs for certain types of information in ads. Although most investigations to determine the appropriate level of information stemmed from consumer needs, there are requirements mandated from political bodies (Mueller 1991). Political bodies were concerned with issues of deception and false advertisement. The large collection of advertisement studies has compared information content in advertisements across cultures, time, media, and different regulatory regimes (Anderson and Renault 2006). From these studies, information cues have been established. Information cues are defined as categories of information that are potentially useful to consumers (Mueller 1991). Resnik and Stern's (1977) set of cues is the most referenced in research. These 14 cues spawned from the results of a content-driven investigation of 378 television commercials. Other researchers have studied the effects of international television advertisements and suggested additions to the original set of cues. Mueller (1991) contributed two additional cues after her comparison of standardized versus specialized multinational advertisements. In 1997, Taylor et al. discovered information gaps in the foundational 14 cues and extended the list to 30 informational cues. These three main constructs are illustrated in Figure 2.4.

Advertisement research is one of the few areas that emphasize the value of information during design. Walker (2002) investigated online advertisements, which first appeared in 1994. The informational and emotional content of Internet advertisements were examined and a set of informational cues, which were all derivatives from Resnik and Stern (1977), were documented. The transferability of the informational cues from television to the Internet was demonstrated by Walker's (2002) research. This also illustrates the long-standing validation of the original set of information cues, which provides a strong foundation for the development of a definitional framework of useful content for Web sites.

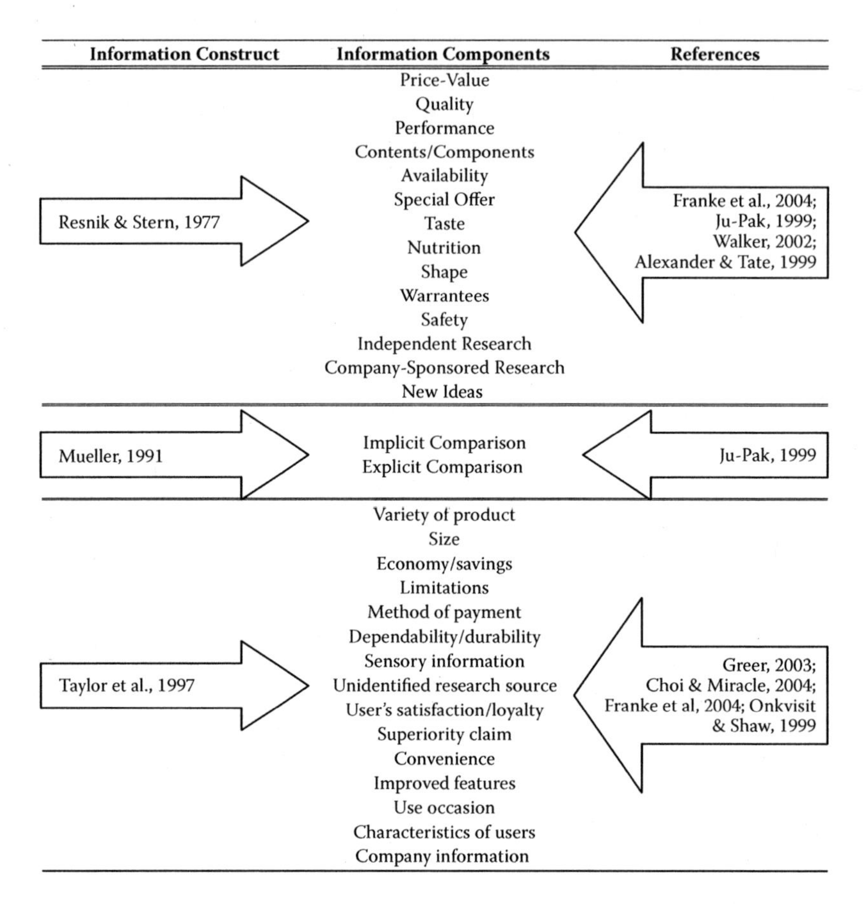

FIGURE 2.4 Essential information components in advertising domain. (Modified from Savoy, A., and G. Salvendy, 2008. Foundations of content preparation for the Web. *Theoretical Issues in Ergonomics Science* 9(6): 501–21.)

2.2.1.3 E-Education

There are a great number of online universities today, but universities are recognized as the latecomers to the online migration. Only 60% of colleges and universities in the United States had active home pages in 1999 (Griffin 1999). After 9/11, universities witnessed a dramatic decrease in campus visits, which warranted new methods of recruitment and influenced the onset of e-recruitment techniques and tools. Among those tools, Web sites were developed offering an impressive set of functionalities: virtual tours, selection of a university or course, retrieval of personal records, and bill payment. Despite e-education's newcomer status, it is already ranked in the top two most popular e-environments.

As a recruitment tool, prospective students view university Web sites for information to assist in their school selection process. Thus, effective content should deliver information that would market the school appropriately (Griffin 1999). There was a survey of 489 institutions' Web sites to review the type of information displayed. Results revealed that the information commonly depicted among the institutions were classified into eight categories (general information about the university; general academic information; admissions information; information for alumnae; information about campus activities or calendar of events; library access; information from the public affairs/relations office; and specific departmental, school, or college

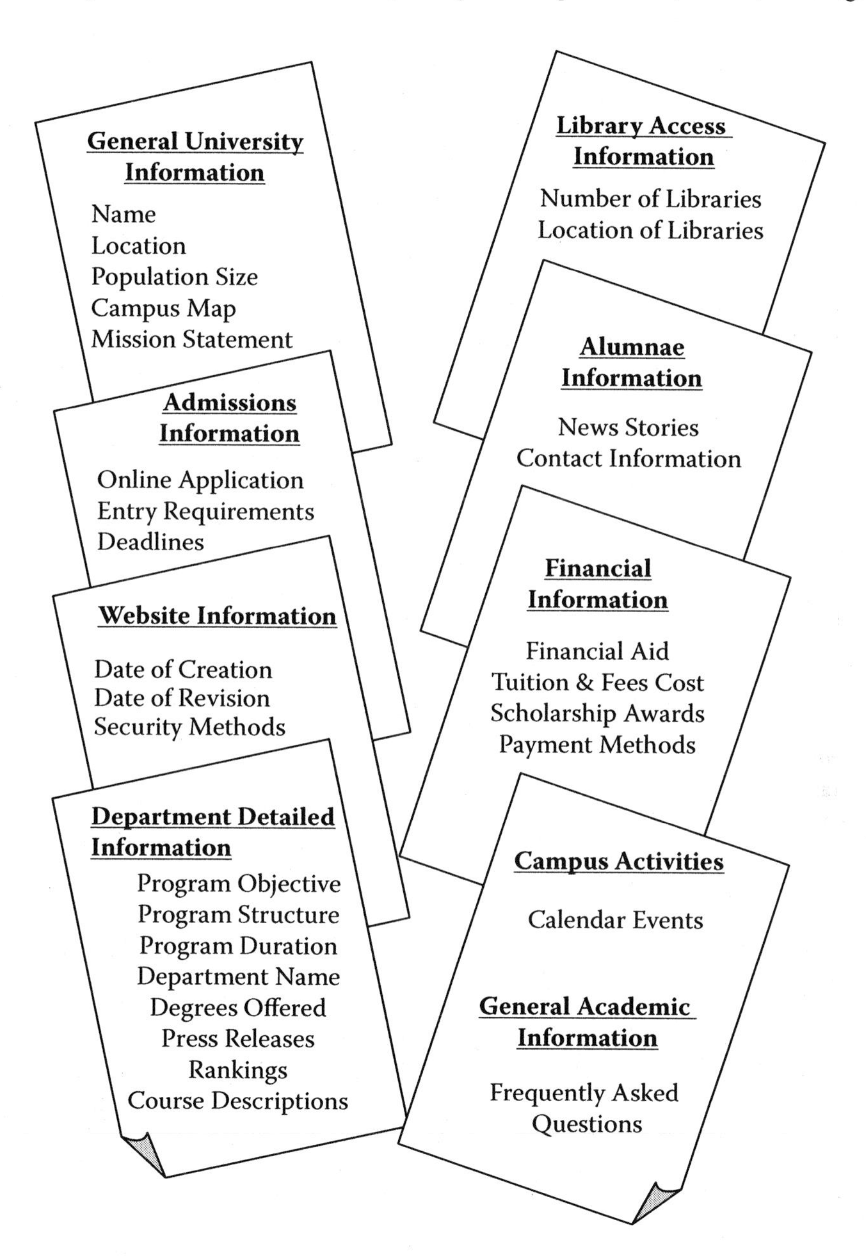

FIGURE 2.5 Essential information components in education domain. (Modified from Savoy, A., and G. Salvendy, 2008. Foundations of content preparation for the Web. *Theoretical Issues in Ergonomics Science* 9(6): 501–21.)

information) (Figure 2.5). The general university information seems to describe basic information elements for this type of Web site. However, those elements of information are crucial to helping students decide if a particular academic environment is the right choice. The information illustrated in Figure 2.5 covers aspects from the admissions process to department specifics. Alumnae information usually arises in conversations or events geared toward fundraising; yet, prospective students are interested in the success of recent graduates and factor that into their selection process. The Columbia (1995) survey provided insight on the current set of information offered. Whether the information offered is satisfactory to users is another question. In fact, there are many important questions unanswered regarding the type of information students want and how much detail the Web site should provide (Savoy and Salvendy 2008).

In 2005, Noel-Levitz (2005) conducted a survey that would provide answers to those desired questions. One thousand high school juniors identified which information elements were useful for recruitment. According to Noel-Levitz (2005), 61% of the students stated that university Web sites should provide a large amount of great content and not many bells and whistles. Students seem highly concerned with the following: tuition (88%), online application (86%), campus events (75%), and faculty (70%) information. This decomposition of information needs has a grave impact on effective content preparation for future designs. Notice that the list—important information—from the Noel-Levitz (2005) is shorter than the previous list—available information. This may suggest a ranking of information elements relative to user needs. Furthermore, the difference in the results from the two surveys discussed stresses the significance of information-centered design theories such as content preparation. Availability and accessibility of information does not define the information as important or useful.

2.2.1.4 Emerging E-Environments

Content preparation should be considered for the design of all e-environments. E-commerce and e-advertising are veterans of e-environments. There are some domains (e.g., financial, government, medical, and entertainment) that are attempting to transition to an online/electronic environment. For example, the governments' attempts to become more citizen-friendly has led to the development of their Web sites and forming e-government (Krauss 2003). In addition, e-education is the e-environment that made an efficient transition to the WWW. New domains need to identify and assess the content and functionality necessary to motivate their audience to use these Web sites (Krauss 2003).

Checklists for the development of basic, advocacy, news, informational, and business Web sites revealed that Web sites spanning from different domains may have similar information content attributes. Sure, there will be some information components specific to a particular domain, but other information components could be domain-independent (Savoy and Salvendy 2008). For instance, all domains need credible information. Previously, it was noted that credible information does not always equate to useful information, but useful information should be credible. This implies that content requirements for quality and credibility (e.g., accuracy, currency, security) could serve as domain-independent information needs. Alexander and Tate's (1999) checklists discussed the following:

- Advocacy site's information concerns were associated with authority and objectivity.
- Information concerns for business sites were focused on authority, objectivity, and interactions/transactions with only one question pertaining to the other criteria.
- The informational Web sites were the only domain with concerns spread out approximately equally among all criteria excluding interaction/transaction.
- News sites evaluations seemed to be mainly concerned with currency and coverage.

Again, the focus of these checklists was IQ and not the determination of information needed for tasks or user information satisfaction. Yet, it supports the following idea. Although there may be some specific information elements for each e-environment that are not transferable across domains, the majority of the elements should be domain-independent and useful for various web interface designs.

Similarities pertaining to information elements enable emerging e-environments to learn from the more established e-environments. During the creation of e-education, there were not many clear guides for its transition and design. Researchers turned to e-commerce designs hoping to learn. Chan and Swatman (2002) reviewed the effectiveness of e-commerce methods in the education domain. This review was founded on the application of Ho's (1997) framework (Chan and Swatman 2002) to the Web sites of universities in Australia and Hong Kong Special Administrative Region. Its results yielded 15 different information components for education Web sites, linking information needs of the education domain with those in the e-commerce domain. Again, the demonstration of transferability encourages the development of a definitional framework of useful information for Web sites that is not limited by domain.

Medical, entertainment, and financial domains are attempting to make the transition to online environments, which makes them latecomers, more so than e-education. Research in these areas is currently being conducted. For example, e-healthcare has been the latest buzz, where researchers are investigating electronic medical records and the information that needs to be presented. There is a shift in which developers are now asking what information should be presented on these Web sites. More attention should be dedicated to information-centered designs.

2.2.2 Information Appliances

As we have discussed, there is an imperative need for the incorporation of content preparation in the design process and evaluation of web-based products and interfaces. Content preparation has been studied on web-based products like e-commerce Web sites, and content factors have been found, such as security content, quality content, service content, appearance description, contact information (Guo and Salvendy

2007; Liao et al. 2009). The content and control for non-web-based products, such as the information appliances, would be quite different from that of a web-based product.

An information appliance is a device that focuses on handling a particular type of information and related tasks. The major differences between web-based products and information appliances are categorized by the design-related content and control. Contents of e-commerce Web sites are mainly concerned with describing the selling product, and how to use the Web site, but contents of information appliances are more about the product functions and aid content for those functions. Compared with web-based products set up on the computer interface, it is harder to control the interface of information appliances. Because most information appliances have more than 20 functions, it is hard to have more than 20 buttons on the panel.

Control is a major aspect in the design of information appliances. Users of information appliances usually need to press the control panel or remote control several times when they want to change the setting or go to a certain function. Sometimes they cannot figure out how to operate the control, or they might press the button more times than needed and then have to spend a long time pressing the keys again. A typical example is how users change their television or monitor settings. A similar scenario often happens to mp4 player users, cell phone users, or digital camera users.

Typical information appliances include mp4 players, liquid crystal display (LCD) TVs, cellular phones, digital cameras, stereo systems, and more. Currently, there are three widely used information appliances: MP3 players, LCD TVs, and cell phones. A comparison between computer-based products and information appliances is provided in Figure 2.6. These products are of different levels of complexity based on the amount of information displayed and the number of keys on the control panel. The MP3 player and LCD TV are considered as low complexity products, because they do not have many functions or much information on the screen. The MP3 player represents the product with median quantity of content and low degree of control. The LCD TV remote has more control keys than the MP3 player but fewer functions. The LCD TV remote represents the product with median quantity content and median degree of control. A cell phone is considered as a median complexity

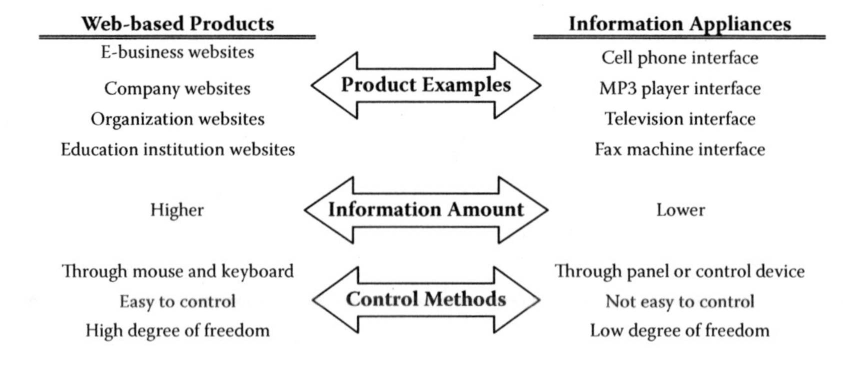

FIGURE 2.6 Comparison between computer-based products and information appliances.

product, because it has a lot more functions and the amount of information is larger than that of an MP3 player and LCD TV. The cell phone represents the product with high quantity of content and median degree of control.

Given some of the similarities between information appliances and web-based products/interfaces, it would not be surprising if they share some information factors. In certain cases, both could define "better" interfaces as those yielding higher customer satisfaction, shorter accomplishment time, and lower error rate. However, it is unlikely that the factors for web-based products would encapsulate all of the factors that pertain to information appliances. The control aspect of these types of products warrants their own investigation.

2.3 CULTURAL DIFFERENCES

In this section, design concepts and theories are presented regarding the Web and information appliance interfaces. Effective design principles and processes include users. Users' preferences, attitudes, and needs are important to design development and evaluation. For instance, usability is important for all interactive interfaces and it is based on user satisfaction. Content preparation is based on the user's information needs for enhanced decision making. Design theories and techniques that lack consideration of the user are not effective; users play an essential part in the design.

Some design decisions are made based on users' cognitive abilities, physical limitations, attitudes, or perceptions. Users can vary by a number of different demographics or characteristics such as age, gender, education, Internet usage, culture. These demographics may have an impact on user preferences and experience, which explains why a design may be great for one user and horrible for others. For example, information needs could be dependent on gender. While females may want solutions to health problems, males may want scores for last night's game. Gender may play a part in design; there are many examples of gender-based aesthetic designs. Just as some designers adjust colors from pink and other pastel colors for females and bold, dark colors for males, the content must be adapted for the respective audiences.

With today's technological advances, many interface designs are gaining international attention (e.g., e-commerce Web sites). Designs that are not limited to country borders warrant research on cultural designs. What is culture? Culture was defined as, "the collective programming of the human mind that distinguishes the member of one human group from those of another. Culture in this sense is a system of collectively held values" (Hofstede 2001, 5). Culture influences human behavior, beliefs, and values, which affect users' attitudes, preferences, and perceptions. With that type of influence, a user's culture could easily influence satisfaction with an interface design; therefore, the broad demographic of culture is important in the design process. From the subculture perspective, the following sections discuss user Internet usage and behavior influenced by attitudes, perceptions, and preferences. Then, a broader perspective is taken as discussions turn to cross-cultural comparisons, where design considerations are influenced by cognitive abilities and styles.

2.3.1 Among American Populations

The United States is home to over 220 million Internet users (Miniwatts Marketing Group 2008). There is no doubt that Americans are avid Internet users, where approximately 73% of the entire population is represented, and there has been a 131% growth of users from 2000 to 2008 (Miniwatts Marketing Group 2008). Even though the United States can be viewed as a collective group when doing cross-cultural comparisons, this group can be subdivided to view the differences in dimensions of Internet usage for the sake of design that will cater to the specific needs of user groups. In the United States, a major division of users or consumer groups is based on race, which can be viewed as a subculture in America. African-Americans and Caucasian-Americans demonstrate unique Internet usage patterns and trends. Regarding usage, there are two things that we are referring to—frequency of use and how the Internet is used. Frequency of use is discussed to describe the rate of Internet use. The latter is an attempt to describe how a group uses the Internet. Regarding, what they do when they are on the Internet, which has direct relation with the different domains we discussed earlier in this chapter.

2.3.1.1 Digital Divide: Impact on Internet Behavior

When technology is discussed in context of African-American and Caucasian-American comparisons, the term *digital divide* will also make its way into the conversation. This term has lingered around the issue of race and technology for years. It is often used to describe the differences of access to technology (e.g., computers, Internet) among racial and ethnic groups in the United States. Commonly, the digital divide pertains to the severely limited access to computers that African-Americans have in comparison to Caucasian-Americans, but it is not limited to only these two groups. More often than not, it is studied as a matter of personal computers (PC) ownership. The digital divide highlights the difference in the number of households within the African-American culture that own a PC versus the number of Caucasian-American households that own a PC.

Researchers have tried to explain why there is such a big difference between African-Americans and Caucasian-Americans with regard to computer access. It has been an issue of interest by the government as well; they are trying to create and install new policies to equal the playing field as computers and technology become more of a life necessity rather than a privilege. They notice that if certain racial groups are affected by the digital divide they will be left behind in the Information Revolution and the consequences would be detrimental. Percentages of the population would be deprived of economic development, electronic recreation, employment opportunities, and more. The technological marginalization of people would have a negative impact on quality of life (Johnson 2008).

Disparities among income, education, and place have been noted for the differences in PC ownership and are driving forces behind the digital divide. Research reported African-Americans and Hispanics tend to be poorer and have lower educational attainment. Consequently, computer ownership is strongly linked to placement and income (Ono and Zavodny 2003). Socioeconomic environments play a role in the access aspect of the digital divide. In other words, placement matters.

Individuals' environments, where they live or the people that live around them, can influence computer ownership (Ono and Zavodny 2003). If one lives among others that own computers and believes in the benefits of its ownership, then one is more likely to own a computer as well. Unfortunately, 2000 consensus data reveal that African-Americans are still disproportionately likely to reside in areas that are primarily segregated and poor (Mossberger et al. 2006). This is influenced by income. The communities of poverty that are often referred to in publications are populated with more African-Americans than Caucasian-Americans. In this placement, the ownership of computers and their benefits are low. Therefore, the need or drive to own one does not exist.

The affects of the digital divide provide critical input for design decisions. Its influences mutated into different user groups' signature Internet use and patterns. Issues of access have led to a user group whose Internet usage occurs primarily on a public computer. This type of access defines a user group with amplified concerns of privacy. Also, this group only has access to the Internet during specified time slots. In addition, there is a group whose Internet usage is limited by connection technologies (e.g., dial-up) and software. This limitation influences the content viewed by users. Digital divide influences different behaviors.

Some may think that access to computers at public libraries, schools, and places of employment are a viable solution to the digital divide. However, this access is not mirrored for a number of African-American individuals when at home. There are lines at public computers. They are public computers, which means that everyone has access to them. People who own their own computers are cautious about the security measures implemented on their computers. Everyone is worried about identity theft. No one wants to enter their passwords or view personal information, files, or records unless they are on their home computer. If that is the case, you definitely do not want to save any information on a public computer. Think about it. All the other places of access have set hours of operation. How many times do you work on your computer after hours? If you did not have access to the Internet all evening, how would the patterns of your Internet usage change? This would definitely affect your usage habits. Well, this is the reality for a number of people. In addition, even though there is access to computers in public libraries, schools, and places of employment, it does not mean one has access or is allowed to access the Web sites or information that is desired. When operating on computers owned by others, one has to follow set rules, which often include stipulations on what Web sites one can visit. Thus, limited access does not solve the problem.

As the price of PCs have decreased over the years, the number of households that own a PC may have increased. However, it seems that some people are always one or more steps behind today's technological developments. Even though access to computers may have improved, access to the Internet could still be a problem. Imagine finally getting a PC; you have scraped your piggy bank and sacrificed to make ends meet to gain ownership. Then, it is revealed that Internet access is not included. There are monthly rates of accessing the Internet. A computer that cannot connect to the Internet is often deemed useless. Accessing the Internet with limited connection speed and software affects the content that users view. In fact, if you have a computer that connects to the Internet with anything slower than broadband, you are missing out on a lot of the online experience. Once you get a computer and get online, there are other technological advances to enhance the user experience. Now the access problem has shifted. Flash and Java allow many Web sites to come alive and provide colors, videos, and animations that are truly remarkable. However, if you try to access those sites with dial-up or digital subscriber line, the experience is not pleasant. Time is spent watching the progression bar more than accessing the information one was seeking. Therefore access could pertain to different dimensions. It is all not captured by the ownership of a PC or the ability to walk to a public library to check your e-mail. This has major implications for online users, e-commerce businesses, and online advertisers.

The authors would be remiss if we did not make it clear that the digital divide does not affect all African-Americans. There were studies conducted in 2008 that clearly state that African-Americans living in more affluent communities are somewhat more likely than similarly situated Caucasian-Americans to have home computer access. It is African-Americans that reside in areas of concentrated poverty that were statistically less likely to have access to a computer. Research shows that African-Americans with incomes below $40,000 were far less likely than Caucasian-Americans to own a computer and go online. However, African-Americans with higher incomes are as likely or more likely to own computers and go online at the same rate as their Caucasian-American counterparts (Mossberger et al. 2006, 612). Despite the digital divide, the number of African-Americans online is increasing. This jump has been credited to women, 61% of the African-American newcomers. There has also been an increase in the number of first-time users (Spooner and Rainie 2000).

2.3.1.2 Subculture Perceptions: Impact of Culture on Internet Behavior

The digital divide does not capture all the differences between African-Americans and Caucasian-Americans. Despite income, the two groups are known to have different cultures. They have different religious beliefs, upbringings, and political perspectives, among other things. These differences also play a part in the usage patterns of the two groups. "Racial differences in both behavior and attitudes are statistically significant even after controlling for differences in income, education, age, and gender, and that this behavior is consistent with broader beliefs about the importance of technology for economic advancement" (Mossberger et al. 2006, 588). As one would expect, these differences would affect how the two racial groups actually use the Internet. Decades of psychological research on how motivational, affective, and cognitive factors influence behavior challenge the assumption that increased access will directly result in use, 'if we build it, they will come' (Jackson et al. 2001, 2020). The online activities of the group vary according to different categories.

Users' perceptions of what type of tool the Internet is will undoubtedly influence their use of the Internet. Two common perspectives view the Internet as either an information tool or a communication tool. African-Americans tend to think of the Internet as an information tool. African-Americans view the Internet as a method of researching different information. They are more likely to search the Internet for information regarding major life issues and religious information. They are more likely than Caucasian-Americans to learn new computer skills in a variety of formats (group instruction, online instruction, printed manuals) and are more willing to use public access sites for computers. African-Americans use the Internet to search for jobs, places to live, and do school-/work-related research. Caucasians are less likely to admit how helpful the Internet is with health care– and hobby-related information. African-Americans are more likely to appreciate the Internet technology and acknowledge its assistance. When asked, 76% of African-Americans agreed that you need computer skills to get ahead compared with only 66% of Caucasian-Americans. Regarding the information tool part, they are less likely to admit that the Internet is good for improving lives.

Caucasian-Americans use the Internet as a communication tool. Interpersonal communication is a major factor. When comparing the two ethnic groups, Caucasian-Americans tend to use e-mail more than African-Americans do. Caucasians are more likely to acknowledge that the Internet helps them stay in contact with their family and friends. The African-American family structure is supported through its connection and communication with the elder members, who do not use the computer as much as the young do. Therefore, the connection between them and their family members is not through the Internet via e-mail as their Caucasian counterparts.

Caucasian-Americans are twice as likely as African-Americans to perform online transactions. African-Americans have less trust in the privacy and confidentiality of e-mail and general web transactions (Jackson et al. 2004). This is due to the demographics belief that the government uses the Internet and WWW to monitor individuals. Therefore, Caucasians use the Internet for making travel reservations, online auctions, and for buying/selling stocks.

Much has to be considered to develop a comprehensive design. Design directives can be derived from aspects associated with subcultures, which provide insight on the specific needs of users. African-Americans and Caucasian-Americans have differences that are based on a number of demographics such as income, beliefs, family structures, location, and upbringing. Table 2.2 summarizes the findings of the few researchers who have conducted surveys to compare African-Americans and Caucasian-Americans with respect to Internet usage. These differences can be used to explain the differences in each group's Internet usage and what types of design would work for one group and not the other or for both.

From a broader perspective, studies have documented that Americans, collectively, share some general concerns regarding Internet usage. First, Americans are concerned with privacy issues when using the Internet. African-Americans are more expressive of the issue than Caucasian-Americans are. Second, Americans are fretful about the content that is available to the children that have access to the Internet. Furthermore, perspectives vary depending on target audiences. In some cases, cross-cultural comparisons are more applicable.

2.3.2 Among American and Chinese Populations

China and the United States are two countries on the forefront of information technology. In the previous section, subculture comparisons were discussed. Direct comparison between two countries offers different insight. Consider the effects of cultural differences on information satisfaction. Cultural differences influence information processing and should affect the design of interfaces. This is exemplified

TABLE 2.2
Cultural Differences among American Populations

Comparison Attributes	Cultural Differences	Groups Compared	References
Access	With respect to the digital divide, Caucasian-Americans have more access to computers and the Internet than African-Americans do.	African-American and Caucasian-American teenagers and adults	Korgen et al. 2001; Ono and Zavodny 2003
Information search	African-Americans use the Internet for information gathering purposes more than Caucasian-Americans do.	African-American and Caucasian-American adults	Spooner et al. 2000
Communication	Caucasian-Americans use the Internet for communication purposes more than African-Americans do.	African-American and Caucasian-American adults	Jackson et al. 2004
Transactions	Caucasian-Americans conduct more business transactions online than African-Americans do.	African-American and Caucasian-American adults	Spooner et al. 2000
Privacy	African-Americans are more concerned with privacy and confidentiality issues associated with online transactions than Caucasian-Americans are.	African-American and Caucasian-American adults	Spooner et al. 2000

best with e-commerce's implementation on a global stage. The status and popularity of e-commerce was discussed earlier. In the international e-commerce setting, companies have the opportunity to retail their products and services to consumers from various countries. This entails interactions involving individuals who may not share common languages, preferences, cognitive styles, or culture. Designers have to consider what design changes must be implemented to accommodate demographic diversity stemming from a cultural context. Can the company use the same Web site design for everyone? Is text translation enough to portray effective global designs? The following sections explore cultural effects on information processing of users, accounting for both normative effects and psychological effects, in the context of e-commerce (see detailed discussion in Liao et al. 2008, 2009).

2.3.2.1 Influences of Cross-Cultural Differences

Socialization sculpts culture, and culture can influence every aspect of people's lives. Genetic traits may be among the most obvious differences in cross-cultural comparisons, but behaviors and thought processes may also differ with this type of comparison. East Asian cultures (e.g., China, Japan, and Korea) and Western cultures (e.g., United Kingdom and United States) have been the focus of much research due to the distinct civilizations that exist on different sides of the world and their roles as fierce competitors and leaders in today's technological advancements (Choong et al. 2005). Cross-cultural comparisons can demonstrate similarities and differences regarding cognitive abilities, cognitive styles, cultural patterns, and infrastructural and economic development. Differences between American and Chinese populations are excellent examples (Table 2.3).

Cognitive abilities pertain to an individual's visual ability, verbal fluency, and digit span. Physiological/biological and societal differences can yield distinct preferences and cognitive abilities between cultures. Discussions pertaining to visual abilities, especially in the context of cross-cultural comparisons, often refer to the orthographical characteristics. Comparing Chinese and American populations' visual abilities is no exception. English is alphabetic language, whereas Chinese is pictographic. This difference has been used to explain why Chinese people possess better visual image processing abilities; Chinese have better visual-form discrimination performance (Carlson 1962; Leifer 1972; Lesser et al. 1965); and Chinese perform better with a pictorial presentation mode of icons over an alphanumeric presentation mode. It is also used to justify why Americans have better performance with alphanumeric presentation mode of icons versus pictorial. Thus, a culture's form/method of written communication affects its visual abilities.

Verbal fluency contributes to differences in cognitive abilities. Chinese children are less fluent verbally than Caucasian-British children are (Lynn et al. 1988). This fluency disparity can be explained by the differences in family order and educational systems that exist between the two cultures. Caucasian-British children are encouraged to express their opinions; however, Chinese children are taught to respect and obey their superiors. In addition, communication patterns (discussed later in this section) of these two cultures help justify this phenomenon.

Digit span was examined as a cognitive ability that varied among cultures. Researchers (Stigler et al. 1986) determined that the Chinese have consistently

TABLE 2.3
Cross-Cultural Differences between American and Chinese Populations

Comparison Attributes	Cultural Differences	Groups Compared	References
Visual ability	Chinese children showed better visual task performance than Caucasian-British children did.	Caucasian-British children Chinese children	Lynn et al. 1988
	Chinese-American children showed better visual task performance than Anglo-American children did.	Anglo-American children Chinese-American children	Jensen and Whang 1994
	Americans performed better with alphanumeric presentation mode of icons than pictorial mode. Chinese performed better with pictorial presentation mode of icons than alphanumeric mode.	Chinese college students US college students	Choong 1996
Verbal fluency	Chinese children were less verbally fluent than Caucasian-British children.	Caucasian-British children Chinese children	Lynn et al. 1988
Digit span	Chinese possessed higher number of mean digit spans than their US counterparts did.	Americans Chinese	Stigler et al. 1986
	Chinese showed superior performance in numerically based tasks than Americans did, but their superiority was attenuated in the older age group.	Americans Chinese	Hedden et al. 2002
Cognitive styles	The cognitive style of US children was inferential-categorical. The cognitive style of Chinese children was relational-contextual.	Chinese children US children	Chiu 1972
	For social events, US children attributed more to internal factors, whereas Chinese children attributed more to external factors.	Chinese children US children	Morris and Peng 1994
	Americans emphasized personal dispositions. Chinese stressed situational and contextual factors.	Chinese college students US college students	Morris and Peng 1994

TABLE 2.3
Cross-Cultural Differences between American and Chinese Populations

Comparison Attributes	Cultural Differences	Groups Compared	References
	Americans performed better when the items were organized by function than by theme. Chinese performed better when the items were organized by theme than by function.	Chinese college students US college students	Choong 1996
	Chinese were more attentive to the relationships and better at detecting covariation in the field than Americans were.	Chinese college students US college students	Ji et al. 2000
	East Asians were more field-dependent than Americans were.	East Asian college students US college students	
	Japanese were more attentive to context and relationships than Americans were.	Japanese college students US college students	Masuda and Nisbett 2001
	East Asians preferred relationship-based categorization. Americans preferred category-based categorization. Mainland and Taiwan Chinese were more relational when tested in Chinese than when tested in English.	East Asian college students US college students	Ji et al. 2004
	Americans fixated sooner and longer on focal objects than the Chinese did. Chinese fixated more on the background than Americans did.	Chinese college students US college students	Chua et al. 2005
Communication context	East Asians prefer a high-context communication pattern and Westerners prefer a low-context communication pattern.		Hall 1983
Cultural dimensions	Cultures vary on five dimensions: power distance, individualism/collectivism, masculinity/femininity, uncertainty avoidance, and long-term orientation		De Mooij 2004; Hofstede 2001

Source: Data from Liao, H., R.W. Proctor, and G. Salvendy. 2008. Content preparation for cross-cultural e-commerce: A review and a model. *Behaviour Information & Technology* 27(1): 43–61

showed a higher number of mean digit spans than their US counterparts. This may be influenced by the Chinese numbers having a shorter pronunciation. Nevertheless, digit span capabilities have also been used to justify consistently higher performance in mathematics by the Chinese (Miller and Stigler 1987; Miuar et al. 1988).

Cognitive styles can be referred to as "stable individual preferences in mode of perceptual organization and conceptual categorization of the external environment" (Kagan et al. 1963, 74). East Asians and Westerners demonstrate differences in attentional patterns, perceptual categorization, and values/norms. East Asians have a tendency to think of elements and their surroundings; they exhibit a relatively holistic attentional pattern. In contrast, Americans' attentional patterns are described as relatively analytical. Americans are inclined to think of an element independent of its context and focus on the decomposition of the element itself and its properties. The perceptual categorization differences of the two cultures seem to support their attentional patterns. In various experiments (Chiu 1972; Unsworth et al. 2005) involving object selection, Chinese participants selected objects based on functional and thematic relationships. Their American counterparts based their decision making on inferences made about the objects that were grouped together (Liao et al. 2008). Apparently, attentional patterns and perceptual categorization are related; the understanding of relationships among elements/objects is derived from attention to fields. Lastly, the two cultures differ with respect to socialization practices, values/norms, and education that shape cognitive styles. Some believe that the differences in cognitive styles stem from the foundation of societal values. Thus, the agricultural history of China and the obedient, cooperative behavior that demanded a successful harvest season is responsible for the well-structured collectivistic network that exists today. Moreover, Westerners' history of autonomous functioning founded in a history of hunting and herding has influenced an individualistic culture with a loose social structure.

In addition, cultural patterns have been explained as a function of communication context and cultural dimensions. Communication context considers the information transmitted directly (verbal) and indirectly (nonverbal). Cultures can be considered high-context or low-context. East Asian nations (e.g., China, Japan, and Korea) are usually considered to be high-context cultures; they tend to adopt indirect verbal communication. In fact, East Asians talk around the point and expect listeners to derive the point from the context of the conversation (Hall 1983). In contrast, low-context communication style is often practiced and preferred by most Westerners (e.g., Americans).

Cultural dimensions have been used to distinguish cultures with regard to their patterns. Hofstede's (2001) model classifies cultures based on five dimensions: power distance, individualism/collectivism, masculinity/femininity, uncertainty avoidance, and long-term orientation. These dimensions are defined as:

1. *Power distance* is the extent to which less powerful members of organizations and institutions accept and expect that power is distributed unequally (Hofstede 2001). It reflects a culture's attitude toward authority (Ji and McNeal 2001). High power-distance cultures tend to exhibit a social hierarchy; low power-distance cultures tend to value equality in rights and opportunity.

2. *Individualism and collectivism* suggests preferences for a loosely knit and tightly knit social structure, respectively. Individualism describes the social structure where individuals take care of themselves and their immediate families only. Moreover, collectivistic individuals expect other in-group persons to care for them (Hofstede 2001). Collectivistic cultures identify with the group, whereas individualists assert their uniqueness.
3. *Uncertainty avoidance* describes people's avoidance of uncertainty and ambiguity because they feel threatened. Cultures with strong uncertainty avoidance have low tolerance for ambiguity in perception. On the other hand, cultures with weak uncertainty avoidance rely on generalist attitudes and common sense (De Mooij 2004).
4. *Masculinity and femininity* refer to roles of genders within a culture. Masculinity cultures place value in achievements and strength. In contrast, femininity cultures emphasize quality of life and selfless service.
5. *Term orientation* is also used to distinguish cultures. Long-term oriented cultures admire thrift and perseverance. However, short-term oriented cultures view the present as more important than preparation for the future (De Mooij 2004). Chinese populations value pragmatism and thrift more than other traditional values.

Lastly, it has been claimed that emerging incomes do not lead to homogeneous consumer behavior (De Mooij 2000, 2003, 2004). This does not mean that the effects of the economic standing of a country can be ignored. Economic standing can determine consumers' purchasing power and product experience. This could also impact consumers' consumption patterns and buying behavior. From country to country, the e-commerce infrastructure varies. Differences could center upon Internet connection methods and/or allowable payment methods. Similar to the access problem described in America with respect to the digital divide, some countries have less access to computers than others. Though this may be the case for rural China, the more developed areas have better access. Furthermore, payment methods are not universal. Not everyone favors the credit card system like the US For example, debit cards are used more frequently in European countries. In China, it is the norm to pay by check, cash on delivery, or wire transfer.

Internet technology is culturally neutral; however, the information content presented on Web sites, especially e-commerce Web sites, are not always culturally neutral. This is evident in the different consumer behaviors compared across cultures when presenting content. It should not be a surprise that different cultural groups need different information. From one culture to the next, perspectives may vary on the same product; certain product characteristics may be more important for one group of people than for another. Furthermore, communication patterns may influence distinct ways of conducting business transactions (McCort and Malhotra 1993). These different aspects demonstrate why global e-commerce is not simply a translation of a Web site from one language to another and adaptation of local transaction systems. It needs to accommodate culture-induced differences in information preferences by employing different marketing communication strategies and providing consumers of different cultural backgrounds with information compatible with their mentalities (Liao et al. 2008). In fact, these aspects warrant customization of information presented on Web sites in all e-environments. Only recently have research initiatives pertaining to cross-cultural web design emerged, which focus on the internationalization and localization of display codes, such features as formats, colors, icons, and graphics (Choong et al. 2005). Now, content preparation enables research on the cultural effects of information needs.

2.4 SUMMARY

In this chapter, aspects of design were discussed. These aspects are building blocks for an information-oriented interface design with culture emphasis. Figure 2.7 illustrates the conceptual model of this proposed design approach. The model is based on the fusion of the two existing design theories—usability and content preparation, which is demonstrated with a solid black line. In addition, it proclaims functionality, information, and aesthetics as the essential interface design components. Dashed arrows in this figure represent direction of influence between the design aspects depicted. In this model, cultural differences have direct influences upon user information needs and the interface design components. Also, user information requires an influence over the interface design components. Furthermore, the model portrays the influence the interface design has on users' decision making/task completion and users' attitudes and satisfaction with their experience. The aspects of decision making and attitude and satisfaction encourage interpretation regarding user evaluation and assessment, which could be used as feedback in iterative process of design.

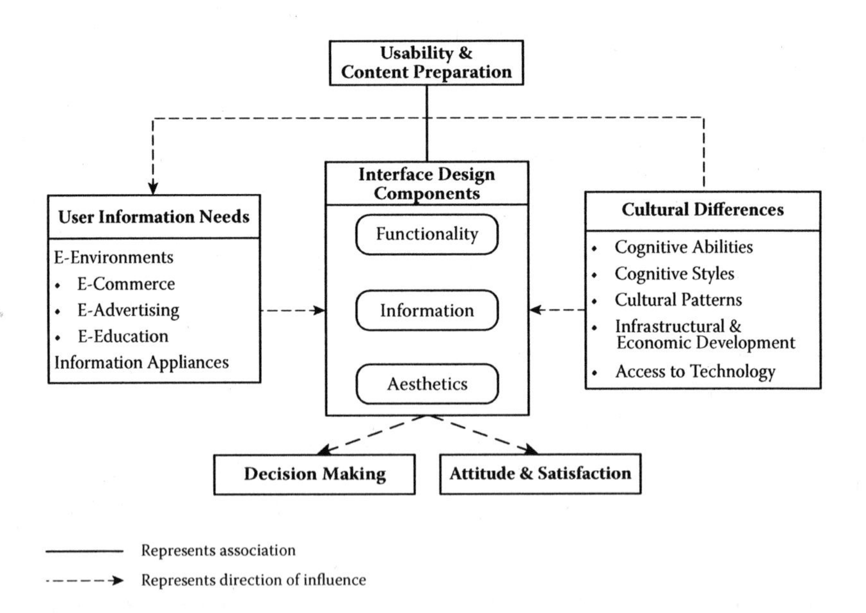

FIGURE 2.7 Information-oriented interface design with culture emphasis. (Modified from Liao, H., R. W. Proctor, and G. Salvendy, 2008. Content preparation for cross-cultural e-commerce: A review and a model. *Behaviour & Information Technology* 27(1): 43–61.)

Usability is always part of discussions and definitions regarding effective interface designs. With the emergence of content preparation, successful interface design has been defined as a function of the how and the what. The how is associated with traditional usability and its emphasis on aesthetics and functionality of interfaces. These aspects are vital to the design of usable interfaces that are efficient, effective, and satisfactory to users. In addition, years of research has provided great insight and guidelines for interface design pertaining to aesthetics and functionality. Moreover, content preparation is defined as an information-oriented design approach. It clarifies the what in good interface design, where emphasis is placed on what information is displayed on the interface. According to Savoy and Salvendy (2008, 1), "the best presentation of the wrong information still results in a design with major usability problems and does not aid the user in accomplishing his task." During the design process, time should be allotted for determining what information is needed for user decision making and completion of tasks. Thus, a design approach that includes usability and content preparation provides a comprehensive approach that covers three main aspects—functionality, information, and aesthetics.

Although most of the research pertaining to interface design addresses usability, there are only a few researchers that focus on information during the design process. Information content is an important component of user interfaces. Information displayed on an interface serves many purposes. It can be used to describe products/services for sale. It can be used as instructions for a device or interface. Overall, information displayed is used for communication. In that, the approach of communication and the information communicated differs with the context. This chapter has reviewed the limited publications that have addressed information components with respect to e-environments and information appliances. User information needs influence interface design, which in turn influences user decision making, attitudes, and satisfaction.

Information content is necessary for users to make rational decisions. Users' information preferences are a function of cultural differences, which consist of differences in cognitive abilities, cognitive styles, cultural patterns, infrastructural and economic development, and access. In the case of HCI, there is normally a human, a computer, and an interaction where each part is as important as the next. The human aspect is often looked upon with regard to the cognitive and physical capabilities. However, social computing has shown that demographics should play a part in design. Interfaces should provide users from different cultural backgrounds with appropriate information content to support decision making. People from different cultures differ in a number of ways such as values, norms, and mentalities. For example, East Asians are more collectivistic and tend to think holistically, whereas Westerners are more individualistic and tend to think analytically. These cultural effects should not be neglected in the design process.

Cultural differences have been shown to have an impact on information needs and design. The significance of research focused on subcultural (within American populations) and cross-cultural (American and Chinese populations) designs is two-fold. Theoretically, it can reveal the underlying mechanisms of how cultural differences influence the information needs of users. Practically, it can provide designers with guidance on how to accommodate information preferences of culturally diverse populations (Liao et al. 2008). The proposed design approach stems from well-established theories including emphasis on culture and information needs specific to types of interfaces. Consideration of the users in the early stages of design should produce interfaces that invoke great user evaluations and overall user experiences.

3 Factor Structure of Content Preparation

3.1 OVERVIEW

Chapter 2 reviews the development of content preparation. Studies of information quality, World Wide Web, advertisement, and other fields have stated essential contents for specific applications. However, there is little experimental data for neither what content is needed by customers nor how the content should be structured. The best way to determine what information application users want is to ask the users themselves. Questionnaires can first be constructed based on the potential content revealed by related literatures, then factor analysis can be conducted on user responses to derive a factor structure, which not only can help IT designers to include all necessary information, but also can be used in other areas.

Because it is not possible to test every application, representative applications should be chosen for questionnaire purposes. We started from applications that need and/or contain large amounts of information. Good representatives are Web sites and information appliances, both of which are information intensive and cover a large range of users. In addition, we wanted to investigate whether the content can serve different user groups in the same way. Therefore we decided to test the content structure on multiple populations.

This goal of this chapter is to derive a general factor structure of content preparation from four studies in this area. Study 1 targeted general Web sites and African- and Caucasian-American populations (Savoy and Salvendy 2008). Study 2 examined the content of e-commerce Web sites needed by the Chinese industrial population (Guo and Salvendy 2009). Study 3 explored the essential content on e-commerce Web sites for portable electronic products with American and Chinese college students (Liao 2008). Study 4 examined the content structure for information appliances used by the Chinese population (Guo et al. forthcoming). Chronologically, Study 3 was first conducted in January 2006, followed by Study 2 in May 2006, Study 1 in January 2007, and then Study 4 in May 2008. As illustrated in Figure 3.1, the scope of the studies expanded chronologically from specific e-commerce Web sites for portable electronic products, to general e-commerce Web sites, then to a larger range of general Web sites. A factor structure of content preparation was derived from each study, and it is our intention to develop a general factor structure for the Web, information appliances, and even broader areas by synthesizing these four factor structures. Furthermore, by comparing the differences in the general factor structure across cultures and applications, we will be able to understand the various influencing factors of the structure (see discussion in Chapter 4).

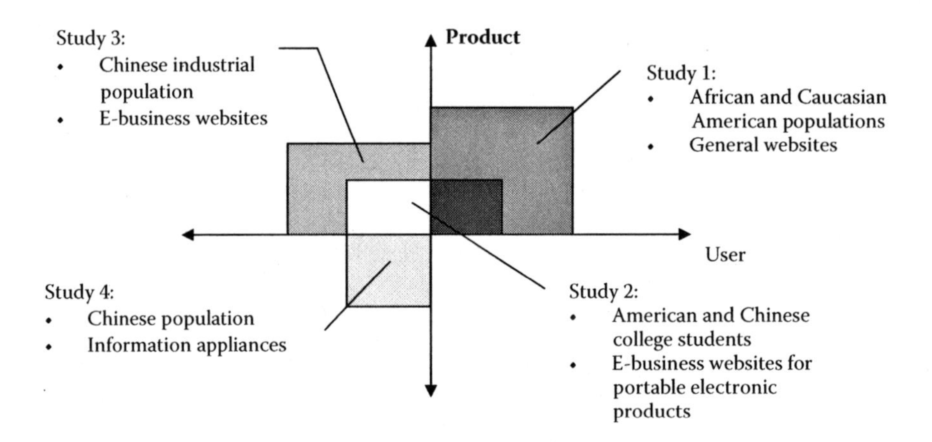

FIGURE 3.1 Connection of four content preparation studies.

Similar experimental procedures were employed in the four studies. The questionnaires were developed from literature about information quality, Web site usability, information appliances usability, and content-related areas such as advertising. Representative application users were surveyed about their opinions on what information they may need in a product or service, or what they may like or dislike. The 7-point Likert scale was used in most of the questionnaires due to its level of sensitivity, detectability, and potential for reducing inter-rater variability. The demographic information of survey participants was also collected. Factor analysis was conducted to identify the hidden factor structure determined by correlations among survey items. Eigenvalues and scree plots were used to determine the number of factors with a factor structure, and factor loadings were used to determine the items (questions) to be included in a factor structure.

3.2 STUDY 1: FACTOR STRUCTURE OF CONTENT PREPARATION BASED ON AMERICAN POPULATION—AFRICAN AND CAUCASIAN AMERICANS

3.2.1 Survey Construction

A 100-item questionnaire (see Appendix A) was constructed based on the review of literature and the following Web sites:

- For e-commerce Web sites, items about ranking, product manual, product accessories, and shopping cart were retrieved from www.amazon.com.
- Date of next update was found on www.nytimes.com. For e-advertising Web sites, information about limitation was found on www.netflix.com.
- Locations and contact information were obtained from www.monster-worldwide.com.
- For e-education Web sites, information about deadlines was retrieved from both www.washington.edu and www.purdue.edu.

The items were grouped into eight hypothetical factors, which are site (3 items), security (2 items), transaction (5 items), shipping (3 items), membership (4 items), company (10 items), customer service (8 items), and product information (46 items). Table 3.1 lists the hypothetical factor structure and the literature supporting each factor and the items within it.

Two scales were used in this study: satisfaction and importance. The *satisfaction scale* was used to measure participants' satisfaction with the information currently presented on the Web sites and was the foundation for refining the measurement tool. For this scale, the 7-point Likert scale was chosen, and a not applicable option was included for items that were not relevant to user tasks or Web sites. The *importance scale* was used to measure participants' value of importance for a particular information item. The response was used in analysis aimed to develop a set of information factors and attributes essential for effective information design. For this scale, a 3-point Likert scale was chosen: important, neutral, and not important.

3.2.2 Procedure and Participants

Data was collected from January to March 2007 with a paper-based survey from 171 participants and a Web-based survey from 129 participants. Participants were instructed to recall the times when they were searching for information on one of the following Web sites: personal interests, work-related, entertainment, e-commerce, or socialization. They were directed to think of particular information that was important during that experience when completing the survey. After they recorded a response for each item, participants were asked to answer the demographic questions listed at the end of the survey.

The diverse group of participants established a good representation of real Internet users. Participants were recruited from universities and communities in more than five states in the US. The collected demographic questions revealed the following respondents' profile (Table 3.2). There were 178 females and 120 males (two participants did not specify their genders), who had been using the Internet for more than one year. Individual ages varied with a mean of 33 years and standard deviation of 12 years. The age group of 20 to 25 years was the largest represented group (27.3% of the participants). Of the participants, 133 were African Americans and 127 were Caucasian Americans with 40 participants of other races. The Web sites evaluated by the participants were from the following categories: 32% from research-related Web sites, 15% from entertainment-related Web sites, 40% from e-commerce Web sites, 8% from socialization-related Web sites, and 5% from school-related and other Web sites. The Web sites used for information gathering (e.g., maps, news) were classified as research-related Web sites. The participants reported a variety of occupations that were classified as academia (27%), students (39%), business (9%), medical (3%), and service (21%).

3.2.3 Factor Analysis

Internal consistency of the survey data was estimated with Cronbach's alpha. There were eight sets of repeated questions included, one set for each hypothesized factor. The overall Cronbach's alpha coefficients were 0.86 and 0.88 for the satisfaction and importance scales, respectively (Table 3.3).

TABLE 3.1
Hypothetical Content Factors and References for Study 1

Factors	Items		References
Site	Creation date, revision date, next revision date		Alexander and Tate 1999; Zhang et al. 1999
Security	Payment security, information privacy		Alexander and Tate 1999; Baierova et al. 2003; Buys and Brown 2004; Kahn et al. 2002; Wang and Strong 1996
Transaction	Confirmation number, payment method, shopping cart, online application, taxes quantity		Buys and Brown 2004; Chan and Swatman 2002; Cheung and Lee 2005; Detlor et al. 2003
Shipping	Shipping cost, tracking number, delivery date		Detlor et al. 2003; Guo and Salvendy in press
Membership	Account status, customization, personal information, entry requirements		Cheung and Lee 2005; Fang and Salvendy 2003
Company	Objective/mission, retailer reputation, retailer service, retailer policy, company/department name, sponsor name, recognition of professional bodies, press releases, employment information, student activities		Alexander and Tate 1999; Chan and Swatman 2002; Detlor et al. 2003; Greer 2003; Katerattanakul and Siau 1999; Klein 2002; Kwon et al. 2002; Wang and Strong 1996; Zhang et al. 1999
Customer service	Help	FAQs, tool tips	Fang and Salvendy 2003; Franke et al. 2004; Gehrke and Turban 1999; Guo and Salvendy in press; Naumann and Rolker 2000; Resnik and Stern 1977; Wang and Strong 1996; Walker 2002
	Contact info	Contact e-mail, contact number, locations	
	Refund Policy	Warranty, return policy, return procedure	
Product information	Components	Product accessories, subjects description	Barnes and Vidgen 2001; Chan and Swatman 2002; Cheung and Lee 2005; Detlor et al. 2003; Fang and Salvendy 2003; Franke et al. 2004; Griffin 1999; Guo and Salvendy 2009; Ju-Pak 1999; Kwon et al. 2002; Muller 1991
	Aesthetics	Color, shape, packaging, size	
	Name	Manufacturer, brand alternative, safety	

TABLE 3.1
Hypothetical Content Factors and References for Study 1

Factors	Items	References
Quality	Evaluation of workmanship, reliability, duration of the program, materials	
Product description	Product compatibility, performance, product technical details, features, convenience, sensory information, program structure, facilities description, nutrition, product availability	
Price	Price value, relative price, discount, scholarship/financial aid, special offer	
Purchase Advice	Independent research, company research, rankings/benchmarking, explicit comparison, customer reviews/alumni	
Other	Variety, product category, product manual, deadlines, limitations/terms of use, new ideas	

Source: Data from Savoy, A. and G. Salvendy. 2008. Foundations of content preparation for the Web. *Theoretical Issues in Ergonomics Science* 9(6): 501–21.

TABLE 3.2
Demographic Information of Participants of Study 1

Variables	Description	Frequency	Percent
Gender	Female	178	59.3
	Male	120	40.0
	Not specified	2	0.7
Education level	High school	29	11.2
	Some undergraduate study	67	25.8
	Undergraduate degree	66	25.4
	Some postgraduate study	35	13.5
	Postgraduate degree	63	24.2
Age group, years	Under 19	11	3.7
	20–25	82	27.3
	26–30	71	23.7
	31–40	61	20.3
	41–60	69	23.0
	Over 60	6	2.0
Web sites	Research	95	31.7
	Entertainment	46	15.3
	E-commerce	120	40.0
	Social	23	7.7
	Other	16	5.3
Job category	Academia	81	27.0
	Business	27	9.0
	Medical	10	3.3
	Service	65	21.7
	Other	117	39.0
Races	African American	133	44.3
	African	1	0.3
	American	4	1.3
	Asian	11	3.7
	Asian Indian	1	0.3
	Caucasian American	127	42.3
	Chinese	2	0.7
	Ethiopian	1	0.3
	European	1	0.3
	German American	1	0.3
	Hispanic	2	0.7
	Indian	2	0.7
	Mulatto	1	0.3
	No response	12	4.0
	Puerto Rican	1	0.3
Survey method	Paper-based survey	171	57
	Web-based survey	129	43

Source: Data from Savoy, A. and G. Salvendy. 2008. Foundations of content preparation for the Web. *Theoretical Issues in Ergonomics Science* 9(6): 501–21.

TABLE 3.3
Internal Consistency of the Survey of Study 1

Factors	Repeated Questions	Satisfaction Scale	Importance Scale
Site	Web site provided current information	0.79	0.78
	I was satisfied with the current information provided		
Transaction	The tax price was precise	0.92	0.85
	The taxes applied to my purchase were precise		
Membership	Web site provided detailed information of my transaction history	0.84	0.77
	The Web site provided detailed information of my transaction history		
Security	The Web site provided a strong sense of security	0.86	0.83
	I felt secure conducting financial transactions on the Web site		
Customer service	Help section provided good instructions for available features	0.80	0.73
	Features of the Web site were supported with good instructions		
Company	Desired information of various company policies was presented	0.72	0.61
	The needed company policies were presented		
Shipping	Information about different shipping methods were helpful	0.88	0.85
	A sufficient amount of information concerning shipment methods		
Product	Components of products/services were explained in detail	0.78	0.70
	Description of the product/services' components was helpful		
Overall		0.86	0.88

Source: Data from Savoy, A. and G. Salvendy. 2008. Foundations of content preparation for the Web. *Theoretical Issues in Ergonomics Science* 9(6): 501–21.

3.2.3.1 Factor Structure Derived from the Satisfaction Scale

Principal factor analysis with varimax rotation was used to identify factors from the satisfaction scale and the loaded items on each factor (Table 3.4). The scree plot, eigenvalues, and percentage of variance explained were used to determine the number of factors. The scree plot displayed a break in the curve at 4, 7, and 12 factors.

TABLE 3.4
Factor Structure of Satisfaction Scale of Study 1

Factors	Items	Mean	Variance Explained (%)	
Factor 1	Product assembly	4.7	4.7	46
General product description	Facility resources	4.6		
	Product nutritional content	4.5		
	Product reviews	5.1		
	Product sensory experience	4.5		
	Product terms of use	4.8		
	Product inventory	4.5		
	Product deadlines	4.6		
	Product comparisons	4.8		
	Independent research of product	4.7		
	Product compatibility	4.9		
	Product limited-time deals	4.8		
	Product purchase advice	5.0		
	Customer testimonies	4.9		
Factor 2	Account status	5.4	5.4	7
Member transaction	Past transactions	5.3		
	Personal information	5.3		
	Overall customer account information	5.4		
	Transaction confirmation	5.8		
	Overall transaction information	5.5		
	Membership requirements	5.3		
	Taxes	5.4		
	Amount of items purchased	5.6		
Factor 3	Product shape	4.9	5.0	5
Shipping	Product size	5.1		
Factor 4	Address	5.4	5.1	4
Secure customer service	Different forms of contact	5.2		
	Instructions for features	5.1		
	Overall customer support	5.2		
	Frequently asked questions	5.1		
	Understood users' problems	4.5		
Factor 5	Company name	5.6	5.3	4
Company	Company reputation	5.2		
	Overall company information	5.2		
Factor 6	Product maintenance	4.6	4.8	3
Durability	New concepts	4.9		
Factor 7	Overall price	4.3	4.3	3
Price	Payment methods	4.3		

Source: Data from Savoy, A. and G. Salvendy. 2008. Foundations of content preparation for the Web. *Theoretical Issues in Ergonomics Science* 9(6): 501–21.

In addition, seven factors could explain 72% of the total variance. Therefore a factor structure with seven factors was chosen. For each factor, items with loadings less than 0.55 were dropped.

The factors derived from the satisfaction scale could be interpreted as general product description, member transaction, shipping, secure customer service, company, durability, and price. The first factor, *general product description,* includes information items that are needed for the Web site to provide a general description of a product or service. For e-commerce Web sites, this factor would include content about assembly of product, nutrition, and sensory experience, among other things. For non–e-commerce Web sites such as company or education Web sites, general product description refers to content items such as product inventory and product deadlines. *Member transaction* refers to aspects of personalized information such as account status, personal information, overall customer account information, membership requirements, and transaction information such as past transactions, transaction confirmation, overall transaction information, taxes, and amount of items purchased. *Shipping* includes information components associated with the shipping process, including methods, delivery dates, and tracking numbers. *Secure customer service* contains information items to ensure a secure environment for customer aid and interactions. It includes six items: address, different forms of contact, instructions for features, overall customer support, frequently asked questions, and understood users' problems. *Company* refers to all kinds of information describing a company and its reputation. *Durability* refers to items describing maintenance as well as the new concept of the product. Finally, *price* refers to prices of products or services and payment methods. Most of the content items summarized by factor analysis are applicable to e-commerce Web sites. However, this result also includes items such as company and membership and is transferable to most types of Web sites.

Although Factor 1 (general product description) explained 46% of the total variance (Table 3.4), the importance of this factor was not ranked first. Instead, Factor 2 (membership transaction) has the highest mean of 5.44, followed by Factor 4 (secured customer service). This ranking reflects some long-standing concerns that customers have: Is it safe to do online shopping? Are my credit cards and privacy information safe? Is there someone that I can consult with if an issue happens? To eliminate these concerns, a Web site should provide not only documents that address Web site security, but also all types of help information (e.g., frequently asked questions) and aid functions (e.g., online chatting for problem solving) to solve customers' security concerns (see discussion in Chapter 5).

3.2.3.2 Factor Structure Derived from the Importance Scale

An additional exploratory factor analysis was performed to establish the factor structure of content information based on responses from the importance scale. Similar to the satisfaction scale, the number of factors was determined by the scree plot and eigenvalues, and percentage of variance was explained. The scree plot displayed a break in the curve at 5 and 9, and the first 12 eigenvalues were greater than 1. Four factors explained 64% of the total variance and eight factors explained 74%. As such, the eight-factor solution was selected based on ease of interpretability.

Principal factor analysis with varimax rotation identified 36 items representing eight factors from the importance scale (Table 3.5). The factors derived from importance values were interpreted as secure customer service, company, member transaction, short product description, product inventory, packaging/shipping, purchase advice, and durability. This factor structure accounted for 76% of the total variance. The importance values highlighted the information that was important to user tasks rather than how well Web sites currently conveyed the information.

3.3 STUDY 2: FACTOR STRUCTURE OF CONTENT PREPARATION FOR E-COMMERCE WEB SITES BASED ON CHINESE INDUSTRIAL POPULATION

3.3.1 Survey Construction

Study 2 aimed for a factor structure that was needed by the Chinese population. According to a report from CNNIC (China Internet Network Information Center) (2006), the first business-to-customer transaction in China was done in March 1998; the first few online transactions happened in the US in 1994. By 2006 (when this survey was conducted), although there were more than 30 million online shopping users in China, the percentage of online shoppers out of all Internet users was only 25% (CNNIC 2006), which was much lower than the 66% penetration rate in the US (Online shopping n.d.). However, the development of e-commerce in China is skyrocketing (Liao et al. in press). Therefore, the survey is of great significance in helping Chinese e-commerce Web sites enhance their content quality.

Through the review of studies listed in Table 3.6, 24 hypothetical factors were proposed. Although most of the factors were supported by background literature, some factors, such as component description, feeling description, operation aid, and match product were added according to the review of e-commerce Web sites in China and the US. The questionnaire derived from the factors is listed in Appendix B. All responses to the questionnaire were scored on a 7-point Likert scale ranging from Strongly Disagree to Strongly Agree, with some items reverse scored to reduce bias. The survey began with general demographic questions, which were followed by randomly ordered items from the e-commerce Web site content.

3.3.2 Procedure and Participants

The survey was conducted in a consumer electronic product corporation in Xiamen, China. The company has more than 5,000 skilled employees and manufactures a wide variety of quality electronics products, such as LCD TVs, HDTVs, LCD monitors, and other telecommunication products. A total of 456 subjects were initially recruited with the help of the senior director; however, 16 participants were deleted because of incomplete responses, and 12 were deleted due to low reliability, which was estimated by repeated questions. Therefore, only the responses from the remaining 428 participants, which are 93.9% of the original sample, were used for further data analysis. Table 3.7 describes general characteristics of the subjects.

Distribution of the participants was gender balanced. Most of the participants ranged in age from 21 to 40 years old. The majority of the participants had associate

TABLE 3.5
Factor Structure of Importance Scale of Study 1

Factors	Items	Mean	Variance Explained (%)	
Factor 1	Warranty description	2.6	2.6	50
Secure customer service	Implemented security	2.6		
	Address	2.7		
	Refund policy	2.6		
	Rebate procedure	2.5		
	Sense of security	2.7		
	Different forms of contact	2.6		
	Overall shipping information	2.6		
	Frequently asked questions	2.6		
	Possible security issue	2.6		
Factor 2	Company credentials	2.2	2.2	7
Company	Company sponsors	2.0		
	Overall company	2.4		
	Company mission statement	2.3		
	Company employment opportunities	2.3		
	Company press release	2.2		
Factor 3	Account status	2.6	2.6	4
Member transaction	Overall customer account information	2.6		
	Past transactions	2.5		
	Personal information	2.5		
	Transaction confirmation	2.7		
	Membership requirements	2.5		
	Overall transaction information	2.6		
Factor 4	Product features	2.6	2.7	4
Short product description	Overall product description	2.6		
	Current information	2.8		
Factor 5	Product deadlines	2.3	2.3	3
Product inventory	Facility resources	2.3		
Factor 6	Product size	2.5	2.4	3
Shipping	Product shape	2.3		
Factor 7	Independent research of product	2.3	2.4	2
Purchase advice	Product comparisons	2.4		
	Customer testimonies	2.4		
	Product reviews	2.4		
Factor 8	Product maintenance	2.5	2.5	2
Durability	Product average lifetime	2.5		

Source: Data from Savoy, A. and G. Salvendy. 2008. Foundations of content preparation for the Web. *Theoretical Issues in Ergonomics Science* 9(6): 501–21.

TABLE 3.6
Hypothetical Factors for E-Commerce Web Sites of Study 2

Factors		Description	Sources
Content about product	Appearance	Appearance description of the product	Detlor et al. 2003
	Wrap	How the product is wrapped	Detlor et al. 2003
	Size	Detail size of the product	Detlor et al. 2003
	Photo	Photos of the product	Detlor et al. 2003
	Specification	The special feature of the product	Detlor et al. 2003
	Manufacturer	Manufacturer name and contact info	Detlor et al. 2003
	Feeling	Feeling about the description of the product	From observation [a]
	Component	Which components are included	From observation [a]
	Quality content	Quality or durability of the product	Detlor et al. 2003; Resnik and Stern 1977
	Detail description	Narrative description of the product	Cho and Park 2001; Corfu et al. 2003; Reibstein 2002; Resnik and Stern 1977; Szymanski and Hise 2000
	Price information	Price, discount price, discount percentage	Abbott et al. 2000; Berr 2006; Corfu et al. 2003; Detlor et al. 2003; Hilsenrath 2002; Reibstein 2002; Resnik and Stern 1977
Content about Web sites	Comment	Comment left by previous customers	Detlor et al. 2003
	Contact information	Contact information of e-business Web sites	Corfu et al. 2003; Liao et al. in press b
	Search	Multiple search functions	Zhang et al. 1999
	Match product	Which products match the selected product	From observation[a]

TABLE 3.6
Hypothetical Factors for E-Commerce Web Sites of Study 2

Factors	Description	Sources
Review	Reviews from expert	Detlor et al. 2003
News	News about discount or product posted on Web sites or e-mails	Corfu et al. 2003
Shipment	Shipping method, time, and cost	Abbott et al. 2000; Cho and Park 2001; Detlor et al. 2003
Service	Service provided by the Web site	Abbott et al. 2000; Cho and Park 2001; Corfu et al. 2003
Link	Link to the manufacturer or other Web sites	Karagiogoudi et al. 2003
Operation	Information about how to operate when customers are online shopping	From observation[a]
Aid function	Functions that would help users in online shopping or decision making	Karagiorgoudi et al. 2003
Responsibility	Responsibility document of transaction	Detlor et al. 2003; Karagiogoudi et al. 2003
Security	Security and privacy of membership and transaction	Abbott et al. 2000; Detlor et al. 2003; Lightner 2003; Reibstein 2002

[a] These hypothetical items were found from observation of American and Chinese e-commerce Web sites. American Web sites: www.amazon.com, www.ebay.com, www.expedia.com, www.priceline.com, www.macys.com, www.target.com, www.dealsea.com, www.fatwallet.com. Chinese Web sites: www.amazon.cn, www.dangdang.com, www.taobao.com.

Source: Data from Guo, Y., and G. Salvendy, 2009. Factor structure of contenet preparation for e-business Web sites: Results of a survey of 428 industrial employees in P. R. China. *Behaviour & Information Technology* 28(1): 73–86.

TABLE 3.7
Demographic Information of Participants of Study 2

Variables	Description	Frequency	Percent
Gender	Male	249	58.2
	Female	179	41.8
Age group, years	21–30	229	53.5
	31–40	113	26.4
	41–50	57	13.3
	51 and above	29	6.8
Education level	High school	56	13.1
	Associate degree	119	27.8
	Bachelor degree	211	49.3
	Master degree and above	42	9.8
Job category	Managers	151	35.3
	Engineers	176	41.1
	Office staff	29	9.1
	Filed sales and marketing	72	16.8
Purchasing experience[a]	Experienced	172	40.2
	Inexperienced	256	59.8
Internet experience[b]	Experienced	363	84.8
	Inexperienced	65	15.2

[a] Purchasing experience: the experience of finishing buying product online.
[b] Internet experience: the experience of surfing and finding information online.
Source: Data from Guo, Y., and G. Salvendy, 2009. Factor structure of contenet preparation for e-business Web sites: Results of a survey of 428 industrial employees in P. R. China. *Behaviour & Information Technology* 28(1): 73–86.

college degrees, bachelor's degrees or higher, and were in senior positions of the company. Around half of the participants were engineers, and more than one-third of them were in charge of management jobs. Most participants (84.8%) had viewed product information on the Internet and about half of them had made online purchases before the survey. The 65 participants who indicated that they had never searched for merchandise online were also included in the study and considered as potential online shoppers.

3.3.3 Factor Analysis

The survey has acceptable overall internal consistency (Cronbach's alpha coefficient of 0.75), as well as acceptable internal consistency for product description content (Cronbach's alpha coefficient of 0.73) and Web site service content (Cronbach's alpha coefficient of 0.75). Eigenvalues and the scree plot were used to determine the number of factors. By examining the eigenvalues, we found that there were 15 factors that had eigenvalues equal to or greater than 1. Moreover, the scree plot showed that there was an elbow point between 15 factors and 16 factors. Therefore 15 factors, which explained 60% of the total variance, were used in this study (Table 3.8). The number of major aspects of e-commerce content seemed to be large. However, because the correlations among questions were low, it suggested that the questions were relatively independent of each other and the questionnaire covered a large range of necessary content information items on e-commerce Web sites.

Both orthogonal and oblique rotation methods produced equivalent factor loading patterns. Items with loading lower than 0.50 were considered insignificant and eliminated. Only one item, the benefit of product, is not loaded. Table 3.8 provides the mean of each factor as well as the variance explained by it. Factor 1 was named *shopping aid* because the items included are related to: where to buy a product when it is sold out, aid video for operation, links to manufacturer, aid picture for operation, photo with size scale, expert comments, and technical support that can give shoppers more information about whether to buy the product. Similarly, Factor 10, *aid function,* contains several functions such as comparisons across Web sites and retailers that can help shoppers make their purchase decisions.

Factor 2 was named *detail description* and Factor 7 was named *appearance description.* These two factors contain information about product color, component, touch/feeling, material, technical information, ingredient, size/weight/volume, production location, smell and flavor, to various photo descriptions. Factor 3 includes some items about cost of service, time of service, and shipping information; therefore it was named *service.* Factor 4 was named *search function* and includes all necessary search functions for e-ecommerce Web sites.

Factor 5 was named *quality* because the majority of items included relate to quality certificates, expiration dates, manufacturers and their reputations, which are indicators of product quality. Factor 6 *security,* which includes transaction security, privacy, and membership, was identified as the most important factor with the highest score. Factors 8 and 9 include information about comments on and ratings of products and were named *comment* and *review aid,* respectively. Factors 11 through 15 contain fewer content items and were named *price content, customized function, matching product, contact information,* and *product specification,* respectively.

Although Factor 6 *(security)* and Factor 5 *(quality)* explained only 2% and 3% of the total variance, respectively, they received the highest mean scores. This means that most online customers consider them as very important and they should have higher hierarchies than other items in the structure. Therefore, when organizations are making decisions regarding the relative importance of each of the 15 factors, they need to consider the relative ranking of the factors as well as the total sample variance that each factor explains.

3.4 STUDY 3: FACTOR STRUCTURE OF CONTENT PREPARATION FOR E-COMMERCE WEB SITES BASED ON AMERICAN AND CHINESE COLLEGE STUDENTS

3.4.1 Survey Construction

The objective of Study 3 was to construct and validate a model for content preparation regarding American and Chinese online consumers' preferences for the content

TABLE 3.8
Factor Structure of Study 2

Factors	Items	Mean	Variance Explained (%)	
Factor 1	Where to buy the product when it is sold out	5.80	5.67	21.9
Shopping aid	Aid video for operation	5.60		
	Link to the manufacturer	5.57		
	Brand of the product	5.70		
	Printable manual	5.77		
	Aid picture for operation	6.00		
	Photo with size scale	5.88		
	Expert comment	5.13		
	Technique support	5.60		
Factor 2	Color of product	5.37	5.49	9.9
Detail description	Product components	5.32		
	Touch or feeling of product	5.33		
	Product material	5.56		
	Product technical information	5.50		
	Ingredient of product	5.66		
	Product weight or volume	5.68		
	Producing place of product	5.35		
	Product smell or flavor	5.65		
Factor 3	Cost of service	6.22	6.15	4.2
Service	Shipping tracking information	6.05		
	Purchase responsibility	6.25		
	Shipping options	6.03		
	Restriction of discounts	6.14		
	Estimate time of service	6.21		
Factor 4	Search by discount information	5.63	5.77	2.9
Search function	Search by price increment/ decrement	5.82		
	Search by product characteristics	5.56		
	Search by price range	5.96		
	Search by product brand	5.96		
	Similar product	5.69		
Factor 5	Product quality certificate	6.54	6.20	2.7
Quality	Specialty of product	6.00		
	Model type of product	6.16		
	Expiration date of product	6.14		
	Reputation of manufacturer	6.36		
	Manufacturer of product	6.01		
	Product accessories	6.16		

(Continued)

TABLE 3.8
Factor Structure of Study 2 (Continued)

Factors	Items	Mean	Variance Explained (%)	
Factor 6	Safety promise of credit card	6.51	6.42	2.4
Security	Customer privacy	6.44		
	Membership	6.32		
Factor 7	Detail description	5.91	5.98	2.3
Appearance description	Appearance of product	5.86		
	Product size	6.03		
	Product photo from different directions	5.94		
	Pictures of all colors/styles	6.18		
Factor 8	Customer comment	5.60	5.38	2.0
Comment	Ratings given by other customers	5.06		
	Customer forum	5.70		
	Search by customer rating function	5.17		
Factor 9	Customer shopping record	4.83	4.96	1.9
Review aid	Record of viewed product	4.99		
	Photos from other customers	5.05		
Factor 10	Price comparison between Web sites and retailers	5.91	5.83	1.8
Aid function	Size chart	6.00		
	Category picture	5.91		
	News of new arrivals	5.57		
	Match product	5.78		
Factor 11	How much could be saved	5.63	5.65	1.7
Price content	The percentage of savings	5.66		
	Price increment/decrement	5.65		
Factor 12	Search bar	5.90	5.83	1.7
Customized function	Site map	5.86		
	Customization function	5.56		
	Contact information	6.00		
Factor 13	Match product	4.62	4.96	1.6
Matching product	Price and duration of shipment	5.30		
Factor 14	E-mail of news about the Web site	5.51	5.89	1.5
Contact Information	Toll-free number service	6.26		
Factor 15				
Product specification	Characteristic of product	5.77	5.77	1.5

Source: Data from Guo, Y., and G. Salvendy, 2009. Factor structure of content preparation for e-business Web sites: Results of a survey of 428 industrial employees in P. R. China. *Behaviour & Information Technology* 28(1): 73–86.

of e-commerce Web sites for portable electronic products. The study focused on a limited scope and defined information content as the information that was needed to make a rational purchase decision and support an online transaction. Based on the work of Detlor et al. (2003), Resnik and Stern (1977), and Taylor et al. (1997), Study 3 proposed a two-level hierarchical classification of information (Table 3.9) (Liao et al. 2008), which contained 23 information items in total. The first level of the classification grouped information content items into two categories: product-related information and retailer-related information. Each category was then further broken down into several information items at the second level. Liao et al.'s (2008) classification has two advantages when compared with previous ones. First, the two categories can well map onto the two steps in the alternative evaluation model proposed by Maes et al. (1999): product evaluation and retailer evaluation. Second, the two-category structure accounts for the unique characteristics of e-commerce by expanding the information classification system of Resnik and Stern (1977) with information items that are necessary for online shopping. For instance, because online consumers cannot touch or smell the products, providing sensory information (e.g., taste, fragrance) is crucial for certain types of products. In addition, because retailer-related information is not necessary in advertisements, it was not included in Resnik and Stern's information classification, whereas such information was necessary and essential for online consumers to complete an online transaction.

Study 3 chose college students from China and the US as the target based on the discovery of GVU (1998) and Ratchford et al. (2001) that online shoppers in both China and the US tended to be young adults with college degrees and a good knowledge of computer use. The scope of the survey was limited to portable electronic products for three reasons. First, at the time when this study was conducted, portable electronic products such as MP3 players, digital cameras, and laptop computers were representative products sold online in both China and the US (ACNielsen 2005; CNNIC 2006; GVU 1998; iResearch 2004). Second, college students in both countries were familiar with and own one or more portable electronic products. Third, there are many types of portable electronic products of various brands and models, and they are normally perceived as typically durable and socially visible products.

There were three parts of the questionnaire: questions about demographic information, questions about product-related information, and questions about retailer-related information. The participants were asked to indicate their preferences for and opinions about information for purchasing portable electronic products online on a 7-point Likert scale ranging from Strongly Disagree to Strongly Agree. Information items in the information content classification that are not applicable to portable electronic products (e.g., nutrition) were not included in the survey.

3.4.2 Procedure and Participants

American students at Purdue University in Indiana, US, and Chinese students at Tsinghua University in Beijing, China, participated in the survey. The students at these two universities can be viewed as representative online consumers in each cultural group because they had good knowledge of computer use and were familiar

TABLE 3.9
Hypothetical Content of Study 3

Categories	Information Cues	Operational Definitions
Product-related information	Name	Brand name
	Manufacturer	The country in which the product was manufactured
	Price value	The cost of the product
		The product's cost-effective value
		The value-retention capability of the product
		The need-satisfaction capability of the product
	Quality	The product's characteristics that distinguish it from competing products (based on an evaluation of workmanship, engineering, durability, excellence of materials, structural superiority of personnel, attention to detail of special services)
	Performance	What the product does and how well it does it
		What it is designed to do compared with alternative purchases
	Components or contents	The components, accessories, or ingredient/nutrition of the product
	Availability	When the product will be available for purchase
	Guarantees or warrantees	The postsales assurances accompany the product
	Safety	The safety features available on a particular product compared with alternative choices
	Research	Research results comparing the product with other products
	New ideas	New technology used in the product
Advantages offered by the product	Special offers	Which limited-time nonprice deals are available with purchase of the product
	Appearance	Weight, size, color of the product
	Sensory information	Information concerning a sensory information such as taste, fragrance, touch, comfort, styling, or sound
	Nutrition	Specific data given concerning the nutritional content of a particular product
	Material	The material that the product is made of
	Convenience	How easy the product is to use
Retailer-related information	Return policy	Retailer's policy on returns and exchanges
	Delivery	Information on the delivery of online purchases
	Contact	The ways to contact representatives

(Continued)

TABLE 3.9
Hypothetical Content of Study 3 (Continued)

Categories	Information Cues	Operational Definitions
	Payment	The payment methods
	Security	Security information about online transactions
	Privacy	Privacy information about the e-commerce Web site

Source: Data from Detlor, B., S. Sproule, and C. Gupta. 2003. Pre-purchase online information seeking: Search versus browse. *Journal of Electronic Commerce Research* 4(2): 72–84; Resnik, A, and B. L. Stern. 1977. Ananalysis of information contentin television advertising. *Journal of Marketing* 41: 50–53; Taylor, C. R.,G. E. Miracle, and R. D. Wilson. 1997. The impact of information level on the effectiveness of US and Korean television commercials. *Journal of Advertising* 26: 1–18; and Liao, H., R. W. Proctor, and G. Salvendy. 2008. Content preparation for cross-cultural e-commerce; A review and a model. *Behaviour & Information Technology* 27(1): 43–61.

with online shopping and portable electronic products. The survey was conducted only in English because Chinese students at Tsinghua University had a reasonable level of English proficiency. Both undergraduate and graduate students at these two universities were targeted. To make the backgrounds of the students at the two universities comparable, invitations with the link to the Web-based survey were sent out via e-mail mainly to approximately 200 American students in the School of Industrial Engineering of Purdue University and approximately 150 Chinese students in the Department of Industrial Engineering of Tsinghua University in December 2005. To recruit as many participants as possible, the invitation asked the recipients to extend the invitation to other interested students in the same or other departments. By January 10, 2006, a total of 76 American and 72 Chinese students responded. The response rates of the American and Chinese engineering students were approximately 30% and 36% of the original number of mailings, respectively. Data of one American student and four Chinese students was eliminated from further analysis due to low internal consistency.

Table 3.10 summarizes the demographic characteristics of the respondents: 45.6% of the 68 Chinese respondents were female, 17.6% were undergraduate students, and 80.9% were engineering students; 37.3% of the 75 American respondents were female, 54.7% were undergraduate students, and 80.0% were engineering majors. The majority of the respondents were engineering students because the massive invitation e-mails were sent mostly to engineering students. The average ages of the American and Chinese respondents were 23.5 years with a standard deviation of 3.94, and 24.5 years with a standard deviation of 2.53, respectively. There was no significant difference between the two groups (t_{141} =1.72, $p = 0.0875$). Thus, the composition of the samples was similar in terms of gender, age, and educational backgrounds, except for a relatively larger portion of the Chinese sample being graduate students.

TABLE 3.10
Demographic Profiles of American and Chinese Respondents in Study 3

		American (n = 75)		Chinese (n = 68)	
Variables	**Description**	**Frequency**	**Percent**	**Frequency**	**Percent**
Gender	Female	28	37.3	31	45.6
	Male	47	62.7	47	54.4
Education level	Undergraduate	41	54.7	12	17.6
	Graduate	34	45.3	56	82.4
Major	Engineering	60	80.0	55	80.9
	Science	5	6.7	3	4.4
	Humanities	6	8.0	6	8.8
	Other	4	5.3	4	5.9

Source: Data from Liao, H., R. W. Proctor, and G. Salvendy. 2009. Chinese and US online consumers' preferences for content of e-commerce Web sites: A survey. *Theoretical Issues in Ergonomics Science* 10: 19–42.

3.4.3 Factor Analysis

The survey contained 33 questions including three sets of paired questions to test the internal consistency of participants' responses. The overall internal consistency as estimated by Cronbach's alpha for the American and Chinese respondents separately and respondents pooled together was 0.92, 0.83, and 0.90, respectively, which indicated that the survey had good internal consistency. Because the two questions in each pair asked the same question, only one of them was included in the data analysis of Study 3. In the present study, principal components factor analysis with varimax rotation was conducted to explore the hidden factor structure determined by the correlations among survey items. Seven factors with eigenvalues larger than 1 were retained and explained 61.5% of the total variance. Although this amount is not considered to be high, it is comparable to other studies involving factor analysis of survey of e-commerce such as Study 1 and Study 2 (Guo and Salvendy in press; Savoy and Salvendy 2008) and a study by Hwang et al. (2006).

In the factor structure (Table 3.11), items with factor loadings higher than 0.50 were considered significant. The items loaded on Factor 1 are about how to return a product and the return and refund policies. Thus, Factor 1 is classified as *product return and exchange*. Factor 2 includes items that are related to online transaction environment, such as transaction security, and potential online purchasing risks, such as safety and privacy, and thus is labeled *online transaction risks*. It should be noted that the item of safety features, although it is not related to online transaction environment, can also affect consumers' perceived online purchasing risks. Items in these two factors all belong to retailer-related information. The survey items in the remaining five factors fall in the category of product-related information and address different aspects of product attributes. Factor 3 is named *product cost and performance* as the items of this factor pertain to product price and product performance. Factor 4 is named *manufacturer reputation,* because the loaded items on Factor 4 are related to the country of the

TABLE 3.11
Factor Structure of Study 3

Factors	Items	Mean	Variance Explained (%)	
Factor 1	Return conditions	4.85	4.97	10.3
Product return and exchange	How to return	4.98		
	Refunded	5.08		
Factor 2	Transaction security	5.71	5.13	10.1
Online transaction risks	Privacy	5.18		
	How to contact	4.95		
	Safety features	4.67		
Factor 3	Price	6.14	5.93	10.0
Product cost and performance	Satisfaction-price	6.44		
	Performance	6.02		
	Convenience	5.44		
	Cost-effectiveness	5.60		
Factor 4	Country	3.50	4.42	10.0
Manufacturer reputation	Value-retention	4.63		
	Warranties	4.82		
	Postsale service	4.72		
Factor 5	Weight	5.25	5.26	8.3
Product appearance	Size	5.87		
	Material	4.66		
Factor 6	New technology	5.53	4.87	6.6
Technology	Manufacturing skills	3.99		
	Quality	5.09		
Factor 7	Composition	4.61	4.76	6.2
Product facts	Research results	4.91		

Source: Data from Liao, H., R. W. Proctor, and G. Salvendy. 2009. Chinese and US online consumers' preferences for content of e-commerce Web sites: A survey. *Theoretical Issues in Ergonomics Science* 10: 19–42.

manufacturer, warranties, and postsales service. Factor 5 is named *product appearance,* which contains items of product size, weight, and material. Factor 6 is named *technology,* because it involves information about the technological specifications of a product and the technology used to manufacture a product. Factor 7 is named *product facts,* as it involves items pertaining to product composition and research results on a product.

3.5 STUDY 4: FACTOR STRUCTURE OF CONTENT PREPARATION FOR INFORMATION APPLIANCES BASED ON CHINESE POPULATION

3.5.1 Survey Construction

Comparisons of these three studies reveal similar factors of content and prove that the quality of content does play an important role in usability. Therefore, it is necessary to explore the proper content and its structure for non-Web-based products. The content for non-Web-based products, such as information appliances, may be different from that for Web-based products. An information appliance is a device that focuses on handling a particular type of information and related tasks. Typical information appliances include music players, cell phones, GPSs, digital cameras, and stereo systems. Content for e-commerce Web sites are mainly concerned with describing products and how to use the Web sites, while content for information appliances is more about the product functions and aid content for those functions.

Cell phones were chosen as the representative information appliance for two reasons. First, nowadays cell phones are being developed as multifunctional devices—which include the functions of digital cameras, music players, GPS devices, and more. Therefore, the factor structure for cell phones could be applied to these appliances. Second, cell phones are one of the most popular information appliances across the world. According to the report from Ministry of Industry and Information Technology of China, there were more than 60 million cell phone users in China by the end of July 2008 ("Statistical monthly report" 2008).

The hypothetical factor structure was developed based on previous studies on content preparation (Guo and Salvendy 2009; Liao et al. in press; Savoy and Salvendy 2008), cell phone usability (Ji et al. 2006; Kaikkonen et al. 2005; Smith-Jackson et al. 2003; Zhang and Adipat 2005), and observation on current advanced cell phones. The hypothetical factors and supporting references are listed in Table 3.12.

Of the seven hypothetical factors, items in *functions* were generally collected from current advanced cell phone features in both China and the US These functions were considered to be part of cell phone content because as more and more functions are added to the cell phone, more information will be needed by users to use these functions. The items of *menu* are text or picture information about cell phone menus, including function name, menu name, function icon, function order, submenu and subfunctions, and scroll bar and cursor. These items were identified as important items by traditional cell phone usability studies (Ji et al. 2006; Zhang and Adipat 2005; Zhang et al. 1999). The *instruction and status* factor includes 15 potential content items that are used for operation instructions or device status stated by cell phone usability studies (Ji et al. 2006; Kaikkonen et al. 2005; Smith-Jackson et al. 2003). The *file* factor contains items for file name, file property, file size, and storage information. These items were introduced to the study because current cell phones are designed to be able to store contact information, text-message, and multimedia files. The *input and search* factor is proposed based on the findings of the importance of "search function" from Study 2 (Guo and Salvendy 2009). Content about input was included because text messaging is widely used among cell phone users of different backgrounds. The service factor was included due to its importance revealed by Studies 1, 2, and 3. The *phone call features* factor was also identified as important by cell phone usability studies (Ji et al. 2006; Zhang and Adipat 2005).

A questionnaire (see Appendix D) was developed based on the factor structure discussed above. This questionnaire included 7 demographic questions and 68 questions about cell phone content. Out of the 68 questions, there were four repeated questions to test internal consistency, and four general questions regarding the importance of content preparation and participants' opinions about current cell phone content.

TABLE 3.12
Hypothetical Content Factors and References of Study 4

Factors	Test Items	References
Functions (18 questions)	MP3 function, camera function, sequential shoot camera function, video recorder function, mobile TV function, cell phone games function, calendar function, electronic book function, electronic dictionary function, GPS function, customization function, local traffic information Web browser function, online chatting function, emergency button function, video phone function, handwriting input function	Smith-Jackson et al. 2003 Also based on current advanced cell phone features
Menu (8 questions)	Function name, menu name, function icon, function order, submenu and subfunctions, scroll bar and cursor	Ji et al. 2006; Zhang and Adipat 2005; Zhang et al. 1999
Instructions and status (15 questions)	Status of all functions, current setting of all functions, back/return key, indication of keys, instructions, confirmation keys, calendar status, message box status, multiple time zone clock, animation	Ji et al. 2006; Kaikkonen et al. 2005; Smith-Jackson et al. 2003
File (6 questions)	File name, file property, file size, storage information	Ji et al. 2006
Input and search (8 questions)	Input method, input content, search by name function, search by "pinyin" function, search by number function	Guo and Salvendy in press; Ji et al. 2006; Zhang and Adipat 2005
Service (4 questions)	Manufacturer contact information, service information, help document, signal carrier information,	Guo and Salvendy in press; Liao et al. in press b
Phone call features (5 questions)	Length of calls, call time, detailed information about missing calls	Ji et al. 2006; Smith-Jackson et al. 2003

Source: Data from Guo, Y., and G. Salvendy, 2009. Factor structure of contenet preparation for e-business Web sites: Results of a survey of 428 industrial employees in P. R. China. *Behaviour & Information Technology* 28(1): 73–86.

3.5.2 Procedure and Participants

The survey was conducted in Xiamen, China, in May 2008. There are two reasons to do the survey in China. One reason is that based on the interview with several experienced engineers and designers working in big electronic companies in 2007, there is no standard or algorithm for designing content for the Chinese market. The other reason is that because Study 2 was conducted in the same city 2 years before Study 3, conducting the two studies in the same place can lend better comparability to them. The paper-based questionnaire was used for the convenience of data collection.

Among 401 participants, 375 responses were retained for further data analysis. Table 3.13 describes the general characteristics of the participants. Similar to Study 2,

TABLE 3.13
Demographic Information of Participants in Study 4

Variables	Description	Frequency	Percent
Gender	Female	158	42.1
	Male	217	57.9
Education	High school	6	1.6
	Associate degree	6	1.6
	Undergraduate degree	316	84.3
	Graduate degree	47	12.5
	Others	0	0
Age, years	Under 20	33	8.8
	20–29	315	84.0
	30–39	15	4.0
	Over 40	12	3.2
Occupation	Manager	28	7.5
	Sales	29	7.7
	Civilian	43	11.5
	Engineer	68	18.1
	Technician or worker	22	5.9
	Teacher or professor	41	10.9
	Student	133	35.5
	Others	11	2.9
Experience using different cell phone models	0	7	1.9
	1	111	29.6
	2	137	36.5
	More than 2	120	32.0
Years of experience using a cell phone	0–1	10	2.7
	2–4	195	52.0
	5–8	123	32.8
	>=8	47	12.5

Source: Guo et al. (forthcoming), used with permission.

distribution of the participants was almost gender balanced. Most of the participants (84%) were young adults between 20 and 29 years old, thus the sample population was representative of the young adults in the Chinese population. About 96.4% of the participants had education higher than associate college. Most of them had experience of using cell phones no less than two years and with two or more models.

3.5.3 Factor Analysis

The survey has acceptable overall internal consistency of 0.82, as well as acceptable internal consistency of four pairs of repeated questions that represent four hypothetical factors of cell phone content: 0.85, 0.79, 0.80, and 0.75. Maximum likelihood factor analysis with varimax rotation and promax rotation was conducted. By examining the screen plot, we found that there is an elbow point between 9 and 10. Therefore nine factors are used in this study, and they explain 85.5% of the total variance. Items with loading lower than 0.50 were considered insignificant and eliminated.

The factors were named according to the loading items (Table 3.14). Factor 1 includes the following content items: current input method, the input "pinyin" letters, what content has been input, search by name, and search by initial. Therefore it is named *information of input and search.* Factor 2 covers items related to assistant functions: number of each function, name of each function, and all options of each function on any menu, scroll bar, and cursor. Therefore, Factor 2 is named *information of functions.* Items of Factor 3 are all related to the indication of keys or functions, such as indications of back to previous menu, confirm key, and which keys are in use. Therefore, Factor 3 is named *information of operation.* Factor 4 includes the three most widely used multimedia functions—digital camera, sequential shooting camera, video camera—and is named *information of multimedia function.* Factor 5 covers items regarding file size, photo size, file properties, and storage, and is named *information of stored files* because all four items are related to cell phone storage space and stored file attributes. Items loaded on Factor 6 are all about phone calls, such as missed call times, time of a missed call, and length of each call, and thus is named *information of phone calls.* Factor 7 is named *help and service information* because the loaded items are about how to get more information about service carrier and manufacturer, as well as help information of cell phone functions. Factor 8 covers a large range of items from reminding icons to emergency key and is considered to be *information of accessorial functions.* Factor 9 is named *information of message* because it contains these two items: icon of message box status and icon of voice mail status.

Table 3.14 also provides the mean of each factor. Factor 6 has the highest mean of 6.14 (higher than the score for Agree), and Factor 4 has the lowest mean of 4.94 (lower than slightly Agree). It means that although Factor 4 explains a little more of the total variance than Factor 6 does, information of phone calls is still more important than information of multimedia functions. Another example is Factor 1, which explains 39.52% of the total variance and has a high mean of 5.90, thus it should be considered a very important factor. Therefore, both variance explained and factor means need to be taken into consideration when constructing the content of a new information appliance.

TABLE 3.14
Factor Structure of Study 4

Factors	Items	Mean	Variance Explained (%)	
Factor 1	Current input method	5.88	5.90	39.5
Information of input and search	The input "pinyin" letters	5.82		
	What is the current input method	5.74		
	What content has been input	5.97		
	Search by name	6.01		
	Search by initial	5.95		
Factor 2	Number of each function	5.01	5.25	10.9
Information of functions	All options of each function on any menu	5.03		
	Scroll bar	5.42		
	Cursor	5.62		
	Name of current function	5.40		
Factor 3	Indication of "back to previous menu" key	5.46	5.33	8.5
Information of operation	Indication of "confirm" key	5.56		
	Indication of current activated key	5.14		
	Indication of which keys are in use	5.14		
Factor 4	Video camera	4.87	4.94	6.8
Information of multimedia function	Sequential-shooting	4.54		
	Digital camera	5.41		
Factor 5	Size of files stored in the cell phone	5.17	5.29	4.8
Information of stored files	Size of photos stored in the cell phone	5.11		
	Properties of files stored in the cell phone	5.19		
	Storage and free space	5.69		

(Continued)

TABLE 3.14
Factor Structure of Study 4 (Continued)

Factors	Items	Mean	Variance Explained (%)	
Factor 6	Missed call times	6.18	6.14	4.4
Information of phone calls	Time of a missed call	6.22		
	Length of each call	6.03		
Factor 7	Contact information of signal carrier	4.97	5.08	3.9
Help and service information	Manufacturer's information	5.06		
	Service information of manufacturer and signal carrier	4.99		
	Help information of each function	5.28		
Factor 8	Icon of "memo" status	5.75	5.87	3.6
Information of accessorial functions	Icon to remind marked days	5.60		
	Emergency key	6.34		
	Name of files stored in the cell phone	5.59		
	Message status	6.05		
Factor 9	Icon of voice message status	5.11	5.16	3.1
Information of messages	Icon of messages box status	5.20		

Source: Guo et al. (forthcoming), used with permission.

Out of the nine factors, four factors are about specific cell phone functions (Factors 4, 6, 8, and 9). These factors and associate items can be applied to the design of cell phone content. The other five factors are related to general functions and operation (Factors 1, 2, 3, 5, and 7). Because these factors are universal, they can be applied to the content design of most information appliances. For instance, Factor 1 can be used for devices that need input functions, such as GPSs and PDAs; Factor 5 is very important for appliances such as music players and digital cameras; and Factors 2, 3, and 7 need to be applied for every information appliance.

The other important information we can obtain from this study is that the input and search factor shows importance by the amount of variance it explains and the factor mean. This factor is essentially important for the Chinese population because it is more difficult to input Chinese characters by using the small cell phone panel and text messaging is more widely used in China.

3.6 GENERAL FACTOR STRUCTURE

When designing content for a certain product, regardless of whether the product is a Web site, an information appliance, or some other product, we need to make sure that users know what the product is and how to use it. Description information of all product aspects can help users to understand what the product is. For example, as illustrated in all four studies we've discussed in previous sections, performance information is essential for the users' decision making. The information about product operation can give customers instructions on how to use the product. For instance, as indicated by all four studies, service information, such as contact information and help information for functions, is necessary for both Web sites and information appliances. As such, a general factor structure is generated by first dividing information content into two broad categories—content of descriptive information and content of operational information—and then dividing the information factors and/or items derived from the four studies into the two categories. The general factor structure was generated and listed in Table 3.15. The *content of descriptive information* includes four main components: sensory description, performance and technology, economic attributes, manufacturer and retailer. The determination of the components was based on the nature of the factors from the four studies. Some factors appear in more than one place due to the new definition of components. The supportive factors and associated items are listed on the right side of the table. *Sensory description* usually covers a wide range of descriptions of users' feelings about the object including the size, weight, color, appearance description, material, component, photos, as well as other properties of products/objects. *Performance and technology* refers to description of objective's performance or product's technology that can help the customer with decision making. It usually includes descriptions of the objective's performance status, quality, specialty, effect, safety information, manual, and instruction. *Economic attributes* refers to any price information of the product that can help the customer with decision making. It usually includes the product price, discount rate, and comparative price to other retailers. *Manufacturer and retailer information* refers to any information about the manufacturer or retailer that can help the customer with decision making. It usually includes the name, introduction, Web site hyperlink, contact

TABLE 3.15
General Factor Structure for the Web and Information Appliances

	General Factors	Study Background	Factors and Items
Content of descriptive information	Sensory description Sensory description refers to description of feeling that users would have over the product. It usually covers a wide range of descriptions, including the size, weight, color, appearance description, material, components, photos, as well as other properties of products/objects.	Study 1 General Web sites American population	Factor 1 (general product description) Product assembly, facility resources, product nutritional content, product terms of use, product deadlines, product comparisons, product sensory experience Factor 3 (shipping) Product shape, product size
		Study 2 E-commerce Web sites Chinese population	Factor 2 (detail description) Color of product, product component, product material, ingredient of product, product weight or volume, touch or feeling of product, product smell or flavor Factor 7 (appearance description) Detail description, appearance of product, product size, product photo from different directions, picture of all colors/styles
		Study 3 E-commerce Web sites for electronic products American and Chinese students	Factor 5 (product appearance) Weight, size, material
	Performance and technology Performance and technology information refers to description of objective's performance or product's technology that would help customers make decisions. It usually includes description of product's performance status, quality, specialty, effect, safety information, manual, and instruction.	Study 1 General Web sites American population	Factor 1 (general product description) product compatibility, product inventory Factor 6 (durability) Product maintenance, new concepts
		Study 2 E-commerce Web sites Chinese population	Factor 2 (detail description) Product technical information Factor 5 (quality) Product quality certificate, specialty of product, product accessories, model type of product, expiration date of product

TABLE 3.15
General Factor Structure for the Web and Information Appliances

General Factors	Study Background	Factors and Items
		Factor 15 (product specification) Characteristic of product
	Study 3 E-commerce Web sites for electronic products American and Chinese students	Factor 3 (product cost and performance) Performance, convenience Factor 6 (technology) New technology, manufacturing skills, quality Factor 7 (product facts) Composition, research results
	Study 4 Information appliances Chinese population	Factor 7 (help and service information) Contact information of signal carrier, manufacturer's information, service information of manufacturer and signal carrier, help information of each function
Economic attributes Economic attributes refers to any price information of the product that would help customers make decisions. It usually includes the product price, discount rate, and comparative price to other retailers	Study 1 General Web sites American population	Factor 7 (price)
	Study 2 E-commerce Web sites Chinese population	Factor 11 (price content) How much could be saved, percentage of savings, price increment or decrement
	Study 3 E-commerce Web sites for electronic products American and Chinese students	Factor 3 (product cost and performance) Price, satisfaction-price, performance, convenience, cost-effectiveness

(Continued)

TABLE 3.15
General Factor Structure for the Web and Information Appliances (Continued)

	General Factors	Study Background	Factors and Items
	Manufacturer and retailer Manufacturer and retailer information refers to any information about the manufacturer or retailer that would help customers make decisions. It usually includes the name, introduction, Web site hyperlink, contact information, location, operating hours, review/evaluation, and reputation of the manufacturer or retailer.	Study 1 General Web sites American population	Factor 5 (company) Company name, company reputation, overall company
		Study 2 E-commerce Web sites Chinese population	Factor 5 (quality) Reputation of manufacturer, Manufacturer of product Factor 14 (contact information) E-mail of news of Web sites, toll-free number service
		Study 3 E-commerce Web sites for electronic products American and Chinese students	Factor 4 (manufacturer reputation) Country, value-retention, postsale service
		Study 4 Information appliances Chinese population	Factor 7 (help and service information) Contact information of signal carrier, manufacturer's information
Content of operational information	Transaction information refers to the content about payment that would help customers make decisions. It usually includes the payment method, security terms and conditions, member privacy guarantee, and membership information.	Study 1 General Web sites American population	Factor 2 (member transaction) personal information, past transactions, membership requirements, taxes, overall transaction information, account status, amount of item purchased Factor 4 (secure customer service) Address, different forms of contact, instructions for features, overall customer support, FAQs Factor 7 (price) Payment methods

TABLE 3.15
General Factor Structure for the Web and Information Appliances

	General Factors	Study Background	Factors and Items
		Study 2 E-commerce Web sites Chinese population	Factor 6 (security) Safety promise of credit card, customer privacy, membership
		Study 3 E-commerce Web sites for electronic products American and Chinese students	Factor 2 (online transaction risks) Transaction security, privacy, how to contact, safety features
Content of operational information	Function Function refers to the listed function or the function that would help customers make decisions, and assistance information for operating the function.	Study 1 General Web sites American population	Factor 1 (general product description) Product reviews, product limited-time deals, product purchase advice, customer testimonies
	For Web sites, it usually includes the functions that would let customers leave their ratings, reviews, and comments of the product they bought,	Study 2 E-commerce Web sites Chinese population	Factor 1 (shopping aid) Where to buy the product when it is sold out, aid video for operation link to the manufacturer, printable manual, aid picture for operation, photo with size scale, expert comment, technique support Factor 4 (search function) Search by discount information, search by price increment or decrement, search by product characteristics, search by price range, search by product brand or similar product

(Continued)

TABLE 3.15
General Factor Structure for the Web and Information Appliances (Continued)

General Factors	Study Background	Factors and Items
		Factor 8 (comment)
		Customer comment, ratings made by other customers, customer forum, search by customer rating function
		Factor 9 (review aid)
		Customer shopping record, record of viewed product, photos from other customers
		Factor 13 (matching product)
		Match product
	Study 4 Information appliances Chinese population	Factor 1 (information of input and search)
		Current input method, the input "pinyin" letters, what is the current input method, what content has been input, search by name, search by initial
		Factor 2 (information of functions)
		Number of each function; all options of each function on any menu, scroll bar, or cursor; name of current function

TABLE 3.15
General Factor Structure for the Web and Information Appliances

General Factors	Study Background	Factors and Items
Service Service refers to the information regarding service provided by the retailer or manufacturer. It usually includes help functions or information about using the product; shipping and shipping options; return, exchange, and warranty conditions.	Study 2 E-commerce Web sites Chinese population	Factor 3 (information of operation) Indication of "back to previous menu" key, indication of "confirm" key, indication of current activated key, indication of which keys are in use
	Study 3 E-commerce Web sites for electronic product American and Chinese students	Factor 4 (information of multimedia function) Video camera, sequential-shooting, digital camera Factor 3 (service) Cost of service, shipping tracking information, purchaser's responsibility, shipping options, restriction of discounts, estimate time of service Factor 12 (customized function) Search bar, site map, customization function, contact information Factor 1 (product return and exchange) Return conditions, how to return, refunds
	Study 4 Information appliances Chinese population	Factor 7 (help and service information) Service information of manufacturer and signal carrier, help information of each function

information, location, operating hours, review/evaluation, and reputation of the manufacturer or retailer. For the second category, transaction, function, and service are included because these factors are used when operating a product. *Transaction information* refers to content about payment especially for Web sites. It usually includes the payment method, security terms and conditions, member privacy guarantee, and membership information. *Function* refers to the listed functions that can help customers make decisions and assistance information for operating the function. For Web sites, it usually includes the functions that allow customers to rate, review, and comment on the product they bought; search by product name, price range, size, color, relevance, best seller, customer review, or new arrival. For information appliances, it covers basic functions to advanced functions. *Service* refers to the information about services provided by the retailer or manufacturer. It usually includes help functions or information for using the product, shipping and shipping options, and return, exchange, and warranty conditions.

The validity of the general factor structure is justified by the fact that each component is supported by one or more factors from at least three studies. Although the general factor structure is largely based on studies related to e-commerce (two of the four studies investigated content usability of e-commerce Web sites), and the general factor structure covers many items used on e-commerce Web sites, the factor structure is transferable to other areas for two reasons.

The first reason is that for Study 1, potential information items were collected from literature on general Internet Web sites, and essential items that are not related to e-commerce Web sites remained if they were loaded in the factor structure, such as facility resources and terms of use. The second reason is that according to Alexander and Tate (1999), who created checklists for the development of basic, advocacy, news, informational, and business Web sites, although function differs across different Internet domains and information appliances, the similarity among the information items across different domains is more than the difference (Alexander and Tate 1999).

Comparison between the factor structure of Web-based products and information appliances show that both consist of three main categories of content: content of descriptive information, content of operation information, and content of service information. The content of operation information for both types of products is similar, and the other two categories differ more or less. It is not surprising to see that no item of Study 4 is included in sensory description, economic attributes, and transaction information, because these factors are only applicable to Web sites. For descriptive information, the factor structure of Web-based products contains information mostly related to the description of the product for sale. Although for information appliances, descriptive information is more diverse, including description of functions to description of status and files. This is due to the device scale differences of large-screen/easy accessed computer, and small-display/narrow keypad cell phone. For service information, Web-based products include more content such as shipping, returning, and transaction, whereas the information appliances have the basic help and service items.

In conclusion, although there may be some specific information items that are not applicable to every domain, the general factor structure, as a starting point, can serve as a theoretical research framework for future studies on content preparation, and can inspire design guidelines to improve content usability in different domains.

4 Cross-Cultural Comparison of Content Preparation

4.1 OVERVIEW

One of the consequences of rapid globalization is the emergence of a new dimension of human factors and ergonomics—cultural ergonomics. The fact that an increasing number of companies begin to market their products and/or services across national, subcultural, and cultural boundaries has called our growing attention to the importance and necessity of this dimension. To better serve users from different cultures, who speak different languages, and who have different economic standings, it is critical to understand the impact of cultural variables on user experience. In this chapter, the significance of cultural ergonomics will be discussed with reference to e-commerce Web site design.

An e-commerce Web site is virtually an information system, thus information content is the prerequisite of its effective operation. Just as subtle facial cues can determine within milliseconds people's automatic and unconscious judgments about the nature of a person's mind and personality, the information cues on an e-commerce Web site send out signals to shape up online shoppers' first opinions about the quality of the site and the trustworthiness of the retailer. It is well-known that first impressions count and are actually so important that they can make the difference between failure, survival, and success. Although we can always revise our first impressions, there are powerful psychological tendencies that prevent us from doing so. As such, content usability has been identified as an important criterion and dimension of overall quality and usability of e-commerce Web sites. The fact that Internet technology is culturally neutral does not necessarily lead to the conclusion that the information content on an e-commerce Web site is also culturally neutral. As discussed in Chapter 2, cultural impact on cognitive content and processes will be reflected in the differences in consumer behaviors. The cultural backgrounds of consumers determine their mental models and product evaluation criteria for a product, and their perception of the importance of a product attribute. The cultural differences may lead to different business transaction models and preferences for a certain communication pattern than others (McCort and Malhotra 1993). One implication of the cross-cultural differences is that the global e-commerce is far more complicated than simply translating a Web site from one language to another and adapting local transaction systems. When it comes to content usability, Web site designers should make every effort to accommodate culture-induced differences in information preferences by employing different marketing communication strategies and providing their Web site users with information compatible with their mentalities (Frascara 2000). Some of those differences are obvious, but others are less clear on the surface. In this chapter, cultural effects on content preparation are discussed with reference mainly to the general factor structure proposed in Chapter 3 (see Table 3.15).

With different cultural origins, China and the US are quite distinct from each other along the dimensions of culture, economy, and e-commerce infrastructure. Therefore, the two nations are compared in this chapter to help us unravel cultural effects on content usability. Additionally, as two typical subcultures within the US, African-American and Caucasian-American cultures are also compared to shed light on how subcultures within a multicultural nation can impact content preparation.

In addition to previous empirical studies, the comparisons are primarily based on the findings of the following studies on content preparation.

- Study 1 described in Section 3.2. The comparative survey results of the 35 items in the purified questionnaire are presented in Table 4.1 for African-American and Caucasian-American populations.
- Study 2 described in Section 3.3.
- Study 3 described in Section 3.4. The comparative survey results of 25 information items are summarized in Table 4.2 for American and Chinese populations.
- Study 4 described in Section 3.5.
- Two experiments by Liao et al. (in press) to study American and Chinese online consumers' information preferences with three particular product categories: MP3 players, digital cameras, and laptop computers. For Experiment 1, two prototype e-commerce Web sites were developed with the same content on both sites but presented in different sequences. The investigators asked 24 American and 24 Chinese university students to purchase the three products on one of the two sites, and their information preferences were assessed and compared in terms of their performance on and attitude toward the sites. For Experiment 2, two versions of the product description were developed for each product used in the experiment with each version focusing on a different information item. The information preferences of 24 American and 24 Chinese university students for a certain information item pertaining to product characteristics were assessed and compared in terms of their preferences for the corresponding product description indicated on a seven-point Likert scale. The experimental results of Experiments 1 and 2 are summarized in Tables 4.3 and 4.4, respectively.

4.2 CROSS-CULTURAL COMPARISONS

4.2.1 Product Descriptive Information

4.2.1.1 Economic Attributes

Products are made to exchange for money, thus it is easy to understand that product prices are one of the indispensable information items on any e-commerce Web site. Undoubtedly, the infinite demand of consumers makes them wish to get the best

TABLE 4.1
Comparisons between African- and Caucasian-American Respondents on Information Items

		African (*n* = 133)			Caucasian (*n* = 126)					
Factors	**Items**	**Rank**	**Mean**	**SD**	**Rank**	**Mean**	**SD**	**% Diff.**[a]	$F_{1,257}$	$p > F$
Sensory	Product size	24	2.55	0.633	19	2.40	0.670	2.53	3.52	0.0616
Performance	Product features*	**8**	**2.69**	0.539	**9**	**2.52**	0.616	2.80	5.47	0.0202
	Overall product description	**14**	**2.65**	0.550	**6**	**2.56**	0.573	1.59	1.86	0.1733
	Product maintenance**	**21**	**2.62**	0.636	28	2.28	0.755	5.65	15.30	0.0001
	Product average lifetime*	22	2.56	0.655	24	2.31	0.721	4.24	8.85	0.0032
	Current information	**2**	**2.76**	0.495	**1**	**2.75**	0.455	0.22	0.05	0.8214
Customer service	Warranty description*	**9**	**2.69**	0.593	**13**	**2.48**	0.713	3.46	6.53	0.0112
	Implemented security procedures	**17**	**2.64**	0.595	**5**	**2.56**	0.651	1.26	0.95	0.3297
	Address	**3**	**2.71**	0.545	**4**	**2.58**	0.611	2.25	3.53	0.0615
	Refund policy	**11**	**2.68**	0.608	**8**	**2.53**	0.666	2.54	3.71	0.0551
	Rebate procedures	**19**	**2.62**	0.658	**16**	**2.46**	0.700	2.73	3.76	0.0535
	Sense of security	**4**	**2.71**	0.516	**2**	**2.67**	0.537	0.79	0.53	0.4673
	Different forms of contact**	**5**	**2.71**	0.501	**10**	**2.50**	0.642	3.57	9.02	0.0029
	Overall shipping information	**18**	**2.64**	0.607	**12**	**2.49**	0.678	2.45	3.39	0.0669
	Frequently asked questions*	**15**	**2.65**	0.605	**14**	**2.48**	0.603	2.84	5.15	0.0241
	Possible security issues	**13**	**2.67**	0.560	**3**	**2.60**	0.609	1.23	1.04	0.3099
Company	Company credentials	33	2.32	0.711	33	2.18	0.720	2.22	2.24	0.1353
	Company sponsors	35	2.13	0.743	35	1.97	0.669	2.66	3.29	0.0709
	Overall company**	25	2.54	0.634	22	2.33	0.645	3.47	6.85	0.0094
	Company mission statement*	30	2.38	0.693	32	2.21	0.719	2.95	4.07	0.0446
	Company employment opportunities	31	2.35	0.769	26	2.30	0.842	0.74	0.20	0.6586
	Company press release	34	2.22	0.732	34	2.17	0.670	0.72	0.25	0.6193

(*Continued*)

TABLE 4.1
Comparisons between African- and Caucasian-American Respondents on Information Items (Continued)

		African (n = 133)			Caucasian (n = 126)					
Factors	**Items**	**Rank**	**Mean**	**SD**	**Rank**	**Mean**	**SD**	**% Diff.**[a]	$F_{1,257}$	$p > F$
Transaction	Account status**	**10**	**2.69**	0.566	**17**	**2.43**	0.662	4.39	11.85	0.0007
	Overall customer account information*	**6**	**2.71**	0.600	**7**	**2.53**	0.641	2.92	5.15	0.0241
	Past transactions**	**12**	**2.68**	0.597	20	2.39	0.704	4.80	12.63	0.0005
	Personal information**	**16**	**2.65**	0.605	**18**	**2.43**	0.662	3.63	7.66	0.0061
	Transaction confirmation**	**1**	**2.77**	0.549	**11**	**2.49**	0.713	4.58	12.16	0.0006
	Membership requirements**	**20**	**2.62**	0.572	21	2.34	0.671	4.71	13.36	0.0003
	Overall transaction information**	**7**	**2.71**	0.574	**15**	**2.47**	0.689	3.98	9.19	0.0027
Purchase advice	Independent research of product*	28	2.43	0.666	30	2.24	0.731	3.17	4.81	0.0292
	Product comparisons*	26	2.51	0.623	25	2.30	0.707	3.49	6.43	0.0118
	Customer testimonies*	27	2.45	0.733	29	2.26	0.706	3.15	4.47	0.0355
	Product reviews**	23	2.56	0.621	23	2.33	0.691	3.85	8.03	0.0050
Product inventory	Product deadlines	32	2.33	0.766	27	2.29	0.847	0.75	0.20	0.6531
	Facility resources*	29	2.41	0.708	31	2.22	0.668	3.06	4.61	0.0327
	Overall mean		2.57			2.40				

Note: Bold items are higher then overall means.

[a] Percent difference of means = {(African-American mean – Caucasian-American mean) / (7 – 1)} × 100.

* $p < 0.05$; ** $p < 0.01$.

Source: Data from Savoy, A., and G. Salvendy. 2008. Foundations of content preparation for the Web. *Theoretical Issues in Ergonomics Science* 9(6): 501–21.

TABLE 4.2
Comparisons between American and Chinese Respondents on Information Items

		American (n = 75)			Chinese (n = 68)					
Factors	**Information Items**	**Rank**	**Mean**	**SD**	**Rank**	**Mean**	**SD**	**% Diff.[a]**	$t_{1, 141}$	$p > t$
Sensory attributes	Color	19	4.45	1.48	22	4.72	1.58	3.33	1.04	0.2988
	Weight	**9**	**5.13**	1.63	**9**	**5.39**	1.43	5.00	1.02	0.3072
	Size	**4**	**6.01**	1.12	**6**	**5.72**	1.08	–5.00	1.59	0.1143
	Material	18	4.51	1.56	21	4.82	1.40	5.00	1.27	0.2059
Product performance	Performance*	**3**	**6.20**	0.84	**5**	**5.82**	1.02	–6.67	2.42	0.0169
	Convenience features	**6**	**5.55**	1.12	**12**	**5.32**	1.21	–3.33	1.14	0.2548
	Quality	**10**	**5.08**	1.34	17	5.10	1.45	0.00	0.10	0.9218
	Value-retention capability**	21	4.16	1.57	16	5.15	1.58	16.50	3.75	0.0003
	Safety features***	23	4.04	1.61	**10**	**5.37**	1.52	23.33	5.06	<0.0001
Technology	New technology used in a product***	**5**	**6.00**	0.97	19	5.01	1.45	–16.67	4.81	<0.0001
	Skills used in manufacturing a product*	24	3.68	1.70	25	4.34	1.64	10.00	2.35	0.0203
Product price	Price**	**2**	**6.39**	0.87	**4**	**5.87**	1.17	–8.33	3.03	0.0029
	Need-satisfaction capability/dollars	**1**	**6.47**	0.70	**1**	**6.41**	0.78	–1.67	0.45	0.6582
	Cost-effectiveness**	**8**	**5.29**	1.24	**2**	**5.94**	0.91	10.00	3.53	0.0006
Manufacturer	Country in which a product was made***	25	2.65	1.35	24	4.43	1.63	28.33	7.10	<0.0001
Composition	Product composition	16	4.61	1.30	23	4.62	1.09	0.00	0.00	0.9830

(Continued)

TABLE 4.2
Comparisons between American and Chinese Respondents on Information Items (Continued)

		American (n = 75)			Chinese (n = 68)					
Factors	**Information Items**	**Rank**	**Mean**	**SD**	**Rank**	**Mean**	**SD**	**% Diff.[a]**	$t_{1,\,141}$	$p > t$
Transaction	Transaction security*	**7**	**5.51**	1.47	**2**	**5.94**	1.03	6.67	2.02	0.0452
	Personal information privacy*	12	4.91	1.69	**7**	**5.47**	1.39	10.00	2.17	0.0321
Customer service	Postsales service***	22	4.11	1.63	**8**	**5.40**	1.08	21.67	5.51	<0.0001
	Warranties**	20	4.39	1.56	**13**	**5.29**	1.20	15.00	3.87	0.0002
	Return conditions	15	4.71	1.50	19	5.01	1.45	5.00	1.24	0.2157
	How to make a return	14	4.76	1.51	15	5.22	1.51	6.67	1.82	0.0714
	Refund	**11**	**4.93**	1.56	14	5.25	1.53	6.67	1.22	0.2236
Purchase advice	Research results on a product	13	4.80	1.49	18	5.03	1.23	3.33	1.00	0.3202
Retailer's information	Contact information**	17	4.57	1.37	**10**	**5.37**	1.26	13.33	3.60	0.0004
	Overall mean		4.92			5.28				

Note: Bold items are higher then overall items.

[a] Percent difference of means = {(Chinese mean – American mean) / (7 – 1)} × 100.

* $p < 0.05$; ** $p < 0.01$; *** $p < 0.0001$.

Source: Data from Liao, H., R. W. Proctor, and G. Salvendy. In press. Content preparation for e-commerce invclving Chinese and US online consumers. *International Journal of Human-Computer Interaction.*

TABLE 4.3
Hypothesis Testing Results on Information Items between American and Chinese Groups

Information Items	American	Chinese	Supported?
New technology	√		Yes
Warranties and postsales assistance		√	Yes
Retailer-related information		√	No
Country		√	Yes
Composition and accessories	√		Yes
Size	√		No
Weight		√	Yes
Color		√	No

Note: √: More emphasis.
Source: Data from Liao, H., R. W. Proctor, and G. Salvendy. In press. Content preparation for e-commerce involving Chinese and US online consumers. *International Journal of Human-Computer Interaction*.

products in the market, but the reality is that the incomes of consumers are limited and "you get what you pay for." Given a budget constraint, consumers get the product that maximizes its marginal utility in terms of satisfaction and desirability. With increasing disposable income, the marginal value of money declines. This implies that wealthier consumers not only are able to afford more expensive products, but also are more willing to increase their consumption of a product or service for more utility (e.g., satisfaction and pleasure). In other words, the same product has a relatively less perceived-value to wealthier consumers.

In Table 4.2, the item of need-satisfaction ranked first and had the lowest standard deviations for both American and Chinese groups. This finding confirms that rational consumer behavior is governed universally by the utility maximizing rule. The implication is that prices are not the most important factor in consumer decisions. When manufacturers lower the prices of their products to induce consumers, they should not compromise the value of their products.

Cost-effectiveness and price are both related to product cost, thus it makes sense to speculate that American and Chinese consumers should show consistent information preferences for these two items. However, the results from Study 3 show otherwise. Compared with the American respondents, the Chinese respondents ranked higher and scored significantly higher on cost-effectiveness ($t_{141} = 3.53$, $p = 0.006$) (see Table 4.2), whereas the American respondents ranked and scored higher ($t_{141} = 3.03$, $p = 0.0029$) (see Table 4.2) than their Chinese counterparts. If we look at each group separately, the American respondents scored much lower on cost-effectiveness (5.29) than price (6.39), whereas the Chinese respondents scored comparatively on these two items. These discrepancies may suggest that compared with Chinese consumers, American consumers are more likely to focus on the cost of a product or whether a product is affordable. This may be explained by the fact that the two nations have

TABLE 4.4
Comparisons on Information Items between American and Chinese Groups

Pairs	Information Items	Product Categories	American (n = 24)	Chinese (n = 24)	$t_{1,229}$	$p > t$
Pair 1	Cost-effectiveness	MP3 players	5.2	5.2	–0.10	0.9200
		Digital cameras	5.3	5.3	0.11	0.9121
		Laptop computers**	4.9	5.5	–3.05	0.0026
	Product performance	MP3 players	5.4	5.1	1.52	0.1309
		Digital cameras*	5.6	5.2	2.51	0.0128
		Laptop computers**	5.2	4.4	3.33	0.0010
Pair 2	Value-retention capability	MP3 players*	5.0	4.6	2.09	0.0377
		Digital cameras*	5.4	4.9	2.29	0.0227
		Laptop computers*	5.5	5.0	2.25	0.0255
	New technology	MP3 players	5.6	5.3	1.84	0.0674
		Digital cameras	5.5	5.2	1.66	0.0973
		Laptop computers	5.0	5.1	–0.64	0.5259
Pair 3	Cost-effectiveness	MP3 players	5.3	5.4	–0.78	0.4344
		Digital cameras	5.4	5.4	–0.06	0.9501
		Laptop computers	5.2	5.5	–1.62	0.1070
	Convenience features	MP3 players*	5.3	4.8	2.32	0.0210
		Digital cameras**	5.1	4.2	3.35	0.0009
		Laptop computers*	4.8	4.2	2.39	0.0177

*$p < 0.05$; **$p < 0.01$.

Source: Data from Liao, H., R. W. Proctor, and G. Salvendy. In press. Content preparation for e-commerce involving Chinese and US online consumers. *International Journal of Human-Computer Interaction*.

different levels of disposable income. In 2006, the median household income in the US was US$48,000 ("Household income in the United States" n.d.); however, the average income in China for the same period was only US$2,025 (The World Bank n.d.). As a result, what seems to be a common commodity to American consumers, such as an MP3 player priced at US$100, may seem to be a luxury to average Chinese consumers. Therefore, American consumers tend to perceive the same product with less value and give its benefits less consideration, whereas Chinese consumers tend to think of the purchase as an important decision. Furthermore, Chinese culture has a tradition for being thrifty, which can be partly illustrated by China having the world's highest savings rate (see discussion in Section 4.2.1.2.1). As such, those two factors make Chinese consumers more utilitarian-oriented than their American counterparts. They focus not only on the affordability of a product, but also on whether they can get the most for their money. The implication is that Chinese consumers may be willing to spend more money on extra product benefits once they are convinced that the extra product benefits are worth the money.

It should be noted that in Liao et al.'s (in press) Experiment 2 (see Table 4.4), the differences between the American and Chinese participants on cost-effectiveness varied with product categories, with significance occurring on laptop computers. This indicates that the preferences of Chinese consumers for cost-effectiveness information may be positively correlated with product prices.

4.2.1.2 Sensory Attributes

The sensory attributes of a product play an important role in the decision making of consumers, not only because they make first impressions that count, but also because they are associated with product quality and linked to brand identification. Under some circumstances, a decision cannot be made or is very difficult to make without sensory attributes. For example, we always want to find out how much it weighs when we buy food, and we always look at the color when we buy clothes. For products such as portable electronic products, some sensory attributes, such as size, weight, and color, are key factors in consumers' decision making, because they are often perceived to be associated with the personality, lifestyle, and social status of the owners of those products. For some products, sensory attributes such as gloss, haze, or texture differentiate luxury products from commodities and can be directly translated into economic value. Additionally, sensory attributes are also often used to denote products' functions. For instance, when we walk into any infant's clothing department in the US and some Western European countries, we can easily find a lot of pink and blue clothing. Although we can find pink and blue clothing for baby girls, it is much more difficult to find pink colors for baby boys.

Most humans rely on the five senses to bridge the gap between the external world and their mental worlds. In consumer behavior, all five of our sensory receptors are employed. That is, we use our eyes to see, ears to hear, mouths to taste, noses to smell, and skin to touch in order to sense a product. However, our abilities to receive sensory inputs from a product are greatly limited in the virtual world of the Internet, where physical contact with a product is impossible and only visual and audio information can be presented so far. This problem is worsened by the low reliability of visual representation on computers because manufacturers of computer graphics cards and/or computer monitors have their own standards, which causes the same color to look different across different computers. Thus, effective representation of products' sensory attributes is a challenge for e-commerce, especially for products whose first impressions are not determined by visual or audio stimuli.

Culture impacts products' sensory attributes in three ways. First, the content of sensory attributes is determined by cultural conventions and/or norms. For example, product weight and size are presented in English units in the US, but they are presented in metric units in many other countries. Second, the meanings of sensory attributes perceived by consumers change across cultures. In our color example, the association of the colors of clothing with the genders of infants does not commonly exist outside of the US and Western European countries. As another example, colors obtain symbolism through cultural references in the culture that consumers grow up in. For instance, though white is traditionally the color of funerals in the East, it is the color of weddings in the West. Third, the same sensory attribute weighs in on consumers' decisions differently across cultures. In the study by Liao et al. (in press), it was hypothesized that for portable electronic products, American online consumers would have stronger preferences for size information than Chinese online consumers would, whereas Chinese online consumers would prefer weight and color information more than American online consumers for the following reasons.

4.2.1.2.1 Size

American online consumers' preferences for size information can be attributed to two factors. The first factor is American online consumers' preferences for new technology (see discussion in Section 4.2.1.4). It is not difficult to understand the role that technology plays in reducing the size of portable electronic products just by looking at the evolutionary history of computers. When the first computer was created in the 1940s, it occupied about 1,800 square feet, used about 18,000 vacuum tubes, and weighed almost 50 tons. Thanks to technological breakthroughs and innovations, the size of computers has been reduced dramatically over the past six decades. Nowadays, most laptop computers available on market at affordable prices are only 2 to 3 inches thick and weigh around 5 pounds. The MacBook Air, is merely 0.76 inches thick, with its thinnest section at 0.16 inches. What is amazing is that unlike other small laptop computers, the MacBook Air has a 13.3-inch screen and a full-sized keyboard. With our dedication to technological revolution, we have every reason to believe that technology will remain a driver for decreased sizes of portable electronic products. Hence, it can be concluded that the size of a portable electronic product is highly correlated with and can reflect the level of technology used in the product.

The second factor is American online consumers' relatively more hedonic-oriented consumption styles compared with the consumption styles of their Chinese counterparts. In addition to offering better portability, more enjoyment, and enhanced experience, portable electronic products of smaller sizes are often perceived to be associated with high social status, and thus can make their owners feel more confident and attractive. However, those hedonic benefits come with high prices. According to Maslow's hierarchy of human needs, once the low levels of biological and physiological needs are satisfied, consumers will be motivated to fulfill their higher levels of psychological needs. Because American consumers have more disposable income than Chinese consumers do, American consumers are able to afford more hedonic benefits and are more willing to buy products that may be perceived as luxury items in China to serve ego and self-actualization needs.

The differences in disposable income between the two nations do not tell the whole story. Another reason for American online consumers' hedonic-oriented consumption styles is their willingness to spend. For decades, Americans have complained that their younger generations have a spending problem. Although it is not our intention to discuss whether we should be concerned with this fact amid today's gloomy economy, a glimpse at the savings rate in the US can shed light on the consumption styles of Americans. Since the Great Depression, 2005 and 2006 marked the first time that the US recorded a negative savings rate in back to back years (Toya 2008), and more than 40% of all women had less than US$500 in the bank (Marks and Scherer n.d.). This means that we are seeing a situation where American consumers are spending every penny they possibly can and even borrow on top of that. In contrast, China has an overall savings rate of nearly 50%, by far the highest in the world (Shiller 2006), and an average personal savings rate of about 30% of household income (Roach 2006). Therefore, the high disposable income of American consumers backed by their willingness to spend makes them more likely than Chinese consumers to focus on products' hedonic attributes, and thus have stronger preferences for size information of portable electronic products.

4.2.1.2.2 Weight

The weight of a product should be highly correlated with its size; therefore given the discussion about product size, it may seem paradoxical to speculate that Chinese online consumers perceive weight information of portable electronic products as more important than American online consumers do. Undoubtedly, lighter products can provide better portability and thus more enjoyment, and the technology utilized in a product can determine its weight, however, the specification is justifiable for two reasons. First, the speculation does not necessarily imply that American online consumers do not perceive weight information as important. The second reason is related to the differences between Asians and Caucasians in their anthropometric characteristics. In general, Caucasians are more anthropometrically robust than Asians. For example, Sampei et al. (2003) compared 194 adolescent girls with 356 Caucasian adolescent girls and found that Japanese girls were shorter and lighter than Caucasian girls. Therefore, heavy portable electronic products can, to a greater extent, impair the experience of Chinese online consumers, especially Chinese female online consumers, than American online consumers. One piece of evidence for Chinese online consumers' preferences for weight information is that many young Chinese females treat their small-sized portable electronic products such as cell phones as body accessories and like to hang them over their necks. This can also be seen in Chinese advertisements for portable electronic products (see

FIGURE 4.1 Chinese cell phone advertisement.

Figure 4.1). Thus, it can be concluded that Chinese online consumers pay relatively more attention to weight information than their American counterparts.

4.2.1.2.3 Color

Color plays a vitally important role in the world in which we live and is considered one of the most useful and powerful design tools. Besides conveying its objective meaning, color can be used to realize many effects by creating strong subjective reactions from people. Psychologically, color can cause responses on a subconscious, visceral, and emotional level. For example, nurses in many American hospitals dress in light blues and pale pinks, which are calming and soothing colors, to relax upset patients. Symbolically, color obtains its subjective meaning through people's internalized personal experience as well as their religious, political, and cultural backgrounds. Though there are commonalities in the meanings of colors around the world, a single color may have totally contrasting meanings across cultures due to cultural effects on color symbolization. This can be well illustrated by the earlier example about the color of white, which is associated with life and purity in the US, whereas it is the traditional color of mourning in Eastern Asia. One implication of cultural effects on color symbolization is that color, as a powerful form of communication, should be used for meaning. For colors that share commonality across cultures, they can be used as standardized symbols, such as red and green as universal traffic signals. In Web site design, it has been suggested to designers to consider blue the safest global color if their Web sites are intended for global audience, because blue has mostly positive meanings throughout the world.

For consumer items, color transforms a colorless object to one that is more aesthetically and visually pleasing and attractive. As discussed before, it can also be used to indicate a product's function. More importantly, color has a great influence on the perception of product quality. As such, color is a critical factor in consumers' decision making, especially for fashion products such as portable electronic products, which send out important signals about their owners, such as lifestyles, social status, and personalities (Business Wire 1999; Nitse et al. 2004). Inadequate or inaccurate color information can cause loss of sales, increase of returns and complaints, and, more importantly, result in degraded consumers' trust and royalty. Empirical evidence from industrial studies have indicated that 30% of online shoppers would not buy a product with ambiguous color information (Business Wire 1999), and more than 50% of online shoppers would not go back to the same online retailer if they receive a product in a color other than what is expected (Imation 2001). Therefore, to facilitate consumers' decision making, products' color information needs to be accurately presented. Considering Web sites in Asian countries (e.g., China and Japan) tend to be more colorful than those in the US (Cyr and Trevor-Smith 2004; Kondratova and Goldfarb 2006), Liao et al. (in press) hypothesized that Chinese online consumers would have stronger preferences for color information than their American counterparts would.

4.2.1.2.4 Hypothesis Test Results

According to Tables 4.3 and 4.4, the hypotheses about size and color information were not supported. For color information, although more Chinese participants in Liao et al.'s (in press) experiment tended to inspect it than the American participants, the differences were not significant. For weight information, the Chinese participants inspected significantly more products than the American participants did, which provided evidence for Chinese online consumers' stronger information preferences for weight information.

4.2.1.3 Product Performance

African-American participants perceived information on product features as more important than Caucasian American participants did ($F_{1,257} = 5.47$, $p = 0.0202$) (see Table 4.1). This is understandable given that African Americans, in general, have lower levels of disposable income. Similarly, because Chinese consumers are less wealthy, more utilitarian-oriented, and perceive higher levels of online shopping risks than American consumers do (see discussion in Section 4.2.2), it is reasonable to speculate that Chinese online consumers would place more emphasis on product performance information than American online consumers would. However, Liao et al. (in press) hypothesized otherwise for the following three reasons. First, product cost and product performance are not equally important. Chinese consumers tend to associate product cost with a relatively higher priority due to their low level of income, and thus are more likely than American consumers to sacrifice product performance for low prices. Second, communication patterns affect consumers' preferences for product performance information. It has been reported that advertisements that clearly articulate product characteristics/features and emphasize practical and functional attributes are more effective in low-context cultures (e.g., the US) than high context cultures (e.g., China, Japan, and South Korea) (De Mooij 2000; Lin 2001; Mueller 1987; Okazaki 2004; Okazaki and Rivas 2002; Taylor et al. 1997). The preferences for low-context communication patterns cause American consumers, to a greater extent, to base product evaluation on product performance. Third, American online consumers are more sophisticated and have relatively more experience with consumer products (Ji and McNeal 2001; Rice and Lu, 1988), which makes them more likely focus on product performance information when making a purchase decision.

Consumers' preferences for product performance information may vary with product prices. According to Table 4.4, the American participants showed significantly stronger preferences than the Chinese participants did for product performance information only for digital cameras and laptop computers. By and large, laptop computers are more expensive than MP3 players and digital cameras, which caused the differences between the two groups for laptop computers to be larger than those for MP3 players and digital cameras.

Convenience features can provide more enjoyment and enhanced experience, thus they are more appealing to American online consumers, who, on average, prefer hedonic-oriented consumption styles. In addition, because more convenience features normally mean higher prices, China's low GDP (gross domestic product) per capita may cause Chinese online consumers to be less willing to pay for hedonic experiences than their American counterparts are. This discussion was confirmed by Liao et al. 2009 survey study (see Table 4.2). Moreover, as shown in Table 4.4, the American participants indicated significantly stronger preferences than the Chinese participants for convenience features consistently across three product categories.

Compared with Caucasian-American participants, African-American and Chinese participants tended to place more emphasis on product durability information. For example, as shown in Table 4.1, African-American participants had stronger preferences than Caucasian-American participants did for product maintenance ($F_{1,257} = 15.30$, $p = 0.0001$) and average lifetime information ($F_{1,257} = 8.85$, $p = 0.0032$), and as shown in Table 4.2, Chinese respondents had stronger preferences than American respondents did for product value-retention capability information ($t_{141} = 3.75$, $p < 0.0003$). Economic standing may be a major contributing factor to the differences. Additionally, the cultural dimension of long-term orientation may be another important contributor to the difference in product value-retention capability between China and the US. As discussed in Chapter 2, China is a long-term-oriented culture and values thrift and perseverance, which partly explains China's high savings rate. In contrast, the US is a relatively low long-term-oriented culture and considers that spending now is more important than saving for tomorrow (De Mooij 2004; Hofstede 2001). As a consequence, Chinese consumers are more likely than their American counterparts are to consider a purchase as a long-term investment, hoping that their investment depreciates at a lower rate and in a longer life span.

However, Table 4.4 reveals that the product descriptions focusing on products' value-retention capability made the American participants more willing to buy the products than the Chinese participants. The discrepancy may be explained by the fact that value-retention capability information was presented to the participants with new technology information. Given that American participants had stronger preferences for new technology information than the Chinese participants (Liao et al. in press a, 2009), the contradiction may be attributed to American participants' preferences for new technology information. Because portable electronic products are high technology products, the technology utilized in them, to a great extent, determines the time period in which they can stay in the mainstream before they are replaced by newer models. Therefore, the value-retention capability of portable electronic products and the new technology utilized in them are, to some extent, correlated; hence consumers' preferences for new technology information may be reflected in their preferences for value-retention capability information.

Government and industry strive to ensure that consumer products are safe; however injuries and even deaths still occur. Compared to the US, consumer product safety has taken on an evergrowing and relatively more important position in China. As revealed in Table 4.2, Chinese respondents indicated significantly higher levels of concerns than their American counterparts did ($t_{141} = 5.06$, $p < 0.0001$) over product safety. This finding is echoed by other studies. For example, it has been reported by Plocher and Zhao (2002) that elderly Chinese people emphasized physical activity for good health and were concerned about physical security. Moreover, similar to the survey of Liao et al. (2009), the study reported that health and safety were perceived as more important than convenience features. Additionally, Ji and McNeal (2001) reported that Chinese children's commercials placed emphasis on health information more frequently than American commercials did. They concluded that besides China's relatively low economic development, China's one-child policy is the main contributor to this phenomenon. Under the one-child policy, the only child of the family gets all the love, blessings, and attention of their both parents and even four grandparents, thus the child's health, which is considered to be critical for the child to be successful in the future, is the most important item in the family's everyday agenda. Another contributing factor for Chinese consumers' preferences for safety features may be their concerns raised by online shopping risks of purchasing low-quality products that may be hazardous to their health.

4.2.1.4 Technology Description

New technology information is one of the most important information items for online consumers' decision making, especially for purchasing high technology products. Table 4.2 suggests that new technology information played a more important role in the decision making of American online consumers than their Chinese counterparts (t_{141} = 4.81, $p < 0.0001$). Furthermore, the attitude of the two cultural groups was opposite. The Chinese respondents indicated that new technology information was not important, whereas the American respondents indicated that new technology information was important and ranked it higher than convenience features and cost-effectiveness.

These survey results are consistent with the two experiments by Liao et al. (in press). The first experiment suggested that American participants' decision-making process was facilitated by new technology information. Moreover, the American participants spent significantly more time inspecting new technology information than the Chinese participants did, and the time counted for a significantly higher percentage of total shopping time. In addition, the American participants inspected new technology information more times. In the second experiment, the American participants indicated that the product descriptions focusing on new technology utilized in products would make them more willing to buy those products than the Chinese participants.

These results seemed to contradict previous studies. For instance, "modernity" and "technology" were identified as two dominant cultural values manifested in Chinese magazine advertisements and Chinese TV commercials and were utilized significantly more frequently than in American TV commercials (Cheng 1994; Cheng and Schweitzer 1996). These findings may seem paradoxical because modernity often reminds us of westernization, and China is normally thought of as a typical Eastern country, which is proud of its long historical heritage and holds on to its traditional Eastern values. Cheng and Schweitzer (1996) argued that it was understandable in context. First, China has been striving to achieve modernization of industry, agriculture, science and technology, and national defense since the "open-door policy" at the end of 1970s by encouraging people to employ advanced technology either developed on its own or from industrialized countries (Beijing Review 1978). As such, Chinese people are open to technology innovations. Second, as young Chinese generations are exposed to and even show a tendency to favor Western lifestyles, Western cultural values appear to conflict while coexisting with Chinese traditional cultural values (Ji and McNeal 2001). Third, Chinese consumers are more easily attracted by "newness" of products than their American counterparts are because they are commercially less sophisticated.

These contradictions mentioned can be explained by the following three reasons. First, although Chinese people are passionately embracing technological innovations at an everincreasing speed to increase the levels of productivity, it may be a different story when it comes to their personal lives. Recall that Western lifestyles conflict while coexisting with Chinese traditional cultural values, thus even though Western cultural values make Chinese consumers become more open to new technology than before, Chinese traditional cultural values still make them more likely than American consumers to view it as an unnecessary luxury. Second, the level of consumption in the economy is largely determined by disposable income. Because China is a relatively less developed and low-income country, Chinese consumers give higher priority to the economy aspects (e.g., cost-effectiveness) of a product than new technology. Third, Americans have relatively more hedonic-oriented lifestyles. Although new technology is considered as a utilitarian product attribute in some studies (e.g., Cheng 1994), it may be perceived as a hedonic-related attribute by American consumers, because advanced technology provides enjoyment and enhanced experience. As such, it can be concluded that new technology information is a more important factor in the decision-making process of American consumers than Chinese consumers.

4.2.1.5 Manufacturer Information

A product has two identities: brand nationality and birthplace, which refers to the country in which the product is made. These two identities used to be the same decades ago, but things have changed in today's global economy. To take advantage of low labor and other product costs, many manufacturers in developed countries, such as the US, are moving more manufacturing facilities to and/or sourcing more products from developing nations, such as China. In a free market, it is common to find a product manufactured in a different country other than its brand nationality. For example, most of the Japanese cars in the American market are built in the US with American labor. Under some circumstances, the identities of a product become obscured and complicated. For instance, brand nationality may become meaningless when we realize that Ford has a huge stake in Mazda and that IBM's PC business now is Chinese. Sometimes, it is difficult to tell where a product is made, when it turns out that the product, such as a Chevrolet Malibu or a Ford Windstar, consists of a number of components and assemblies made in several different countries.

Country information may not be as important as other information items, but two major factors make consumers look closely at labels. The first is patriotism. Many patriotic advocates show their "economic patriotism" by buying products made in their own country. For example, a group of Americans have made it their mission to buy American-made products whenever possible in an effort to keep money within the US and stop countries such as Bangladesh, China, India, and Mexico from stealing their jobs. They even invented bumper stickers that say "Got Patriotism? Buy American" to remind their fellow American citizens that it is their patriotic duty to support their country. On Capitol Hill, following the Buy American Act of 1933, which requires that most flags and goods purchased by the US federal government must contain at least 50% American materials, and the federal law that dictates that American flags purchased by the US federal government for veterans' caskets must be made of only American materials and must be entirely produced in the US, bills have been introduced by some lawmakers in an effort to guarantee that American flags flying over government buildings are completely made in the US. The second factor is that where a product is made can be perceived as an indicator of product quality. For example, American economic patriots claim that American-made items are more likely to be more cost-effective, safer, and higher in quality. One of the reasons usually cited for the US automakers' loss of market share are consumer perceptions that imports are higher in quality (Ransom 2008).

It is interesting to find that American and Chinese online consumers are clearly divided by their stance on country information. Table 4.2 indicates that not only did the Chinese respondents score higher than the American respondents did, but also that the attitudes of the two cultural groups were opposite. With an average response below 4 (neutral), the American respondents disagreed that country information was important. In contrast, the Chinese respondents had an average score above 4 (neutral), indicating a positive attitude. Consistently, Guo and Salvendy (2007) also reported Chinese online consumers' positive attitude toward country information. These findings were further confirmed by the experiments of Liao et al. (in press). As a matter of fact, Chinese consumers' strong preference for country information can be observed on some Chinese e-commerce Web sites, which advertize products' country information in a conspicuous place.

The discrepancies in American and Chinese online consumers' attitudes toward country information can be explained by the following three reasons.

1. The utilitarian-oriented Chinese consumers always try to maximize the utility of their spending to a greater extent than American consumers do. It makes sense to speculate that among the same type of products, a product made in a country with advanced technology, outstanding manufacturing skills, and fine craftsmanship is expected to be high quality. However, high quality comes with high prices. Recall that the behavior of Chinese consumers is driven mainly by the cost-effectiveness of products. Low prices are attractive to them, and yet they may be willing to pay high prices for maximal marginal utility. Their seemingly paradoxical mentality makes them collect as much information as possible to strive for balance between price and quality. Country information is among the information they want to collect to aid their decision making. Furthermore, China is in a phase of its development where companies bombard consumers with vastly more choices than they had even a decade ago. Before they become familiar and mature with new products and loyal to some certain brands, Chinese consumers tend to rely on country information to differentiate products regarding quality.
2. There are differences between the cognitive styles of the two cultures. As mentioned in Chapter 2, compared with the American people, whose relatively analytic and inferential-categorical cognitive style makes them inclined to isolate an object from its environment and focus on its properties, Chinese people have a holistic and relational-contextual cognitive style, which makes them tend to bind an object with its contextual information and focus on the thematic relationships among objects. China's holistic cognitive style can be illustrated by one of the philosophies of Chinese medicine, which looks for a unifying theme through the patient's entire presentation, not just their distinct symptoms. Therefore, there is a well-known Chinese saying—treat your headache with podiatry. The tendency of Chinese consumers to see contexts makes them more likely than American consumers to consider the technology level of the country in which a product was manufactured during product evaluation, and, to a greater extent, tend to believe that the quality is positively correlated with the technology level of the country. In contrast, American consumers tend to base product evaluation directly on its quality and are less likely to associate it with country information.

3. There are differences in power distance between China and the US. China's higher level of power distance make the Chinese people show more respect than the American people do for authority, expertise, and certifications. Consequently, Chinese consumers, to a greater extent, assume that a product manufactured with better expertise and in a country with higher levels of technology has better quality.

4.2.2 Operational Information

4.2.2.1 Transaction Information

Compared with Caucasian-American online consumers, African-American and Chinese online consumers tend to pay more attention to transaction, customer service, and retailer information. As shown in Table 4.2, Chinese respondents reported stronger preferences for warranties, postsales assistance, and retailer information, with significance occurring on several items. Moreover, the scores of the Chinese respondents on warranties, postsales assistance, transaction security, personal information privacy, and how to contact information were above the overall mean. In contrast, the American respondents scored below the overall mean except for transaction security and refund information. It should be noted that although transaction security had the highest score among the retailer-related information items for both cultural groups, it ranked second overall and higher than most product-related information items for the Chinese respondents; in contrast, it ranked seventh overall for the American respondents.

According to the experiment by Liao et al. (in press), warranties and postsales assistance information not only shortened the decision-making process of the Chinese participants, but also was used to compare products more frequently by the Chinese participants than the American participants. However, inconsistent with the survey, the experiment revealed that product-return–related information and privacy information was inspected by the American participants not only more frequently but also longer than by the Chinese participants. Furthermore, the Chinese participants appeared to be less reluctant than the American participants to give out their privacy information. The discrepancies between the survey and experimental results may be attributed to the fact that the Chinese online consumers have less online shopping knowledge and experience. In addition, the Chinese participants' willingness to share their privacy information with the e-commerce companies to receive promotion information may be because the Chinese participants were more interested in the information about special low-price deals than their American counterparts due to China's relatively low economic standing.

Similar patterns of information preferences also exist between the African-American and Caucasian-American participants. As shown in Table 4.1, African-American participants scored higher than their Caucasian counterparts on all the items in the factors of transaction and customer service, with significance occurring on most items. For the factor of retailer information, the African-American participants indicated stronger preferences for information about e-commerce companies and their mission statements.

The abovementioned differences can be explained by the concerns caused by perceived online shopping risk. Perceived online shopping risk is considered to be higher than perceived in-store shopping risk and can be conceptualized as "the subjectively determined expectation of loss" (Forsythe and Shi 2003,869). It is a function of the uncertainty about the potential outcome of a particular transaction and the associated possible unpleasantness. It can be divided into the following four prevalent types of risks (Forsythe and Shi 2003):

1. Financial risk, which is considered to be the most apparent barrier to online shopping, refers to the monetary loss to consumers and the possibility of misuse of consumers' financial information (e.g., identity and credit card information). The Internet may not be as secure as we may think. It is bad enough to lose money, but what can be an endless terrible nightmare is to lose identity and/or bank card information on malicious Web sites. In 2007, over eight million Americans became victims of identity theft with losses of over US$49 billion (Privacy Rights Clearinghouse 2007) to financial institutions. Among the four types of perceived risks, financial risk is the most consistent predictor of online shopping behavior, and has a negative impact on the intent of consumers to purchase online, the amount of online transactions, and the amount of money spent online. In other words, increased financial uncertainty leads to increased concern over the security of the Internet, and then results in decreased online shopping activities.
2. Product performance risk refers to the loss caused by the dissatisfying performance of products and is the most frequently cited reason for not purchasing online. As a commercial medium, the Internet changes the way we shop by offering a fundamentally different environment for retailing than traditional media do. Like everything else, the new retailing environment has its advantages as well as disadvantages. One of the inherent drawbacks is that consumers' ability to judge the quality of products is limited on the Internet. They are not able to touch, feel, smell, and try products. Furthermore, product information (e.g., colors) may be inaccurately and/or insufficiently presented. As a result, online shoppers are often apprehensive as to how well the products can meet their expectations.
3. Psychological risk refers to the disappointment, frustration, and shame experienced caused by the disclosure of consumers' personal information. It is quite annoying when your privacy is invaded by telemarketing calls and junk mail. However, those are just privacy violations; what can be worse is that you can lose your identity by giving out your personal information. The painful psychological torture deters people from initiating an online transaction.
4. Time/convenience risk refers to the loss of time and inconvenience incurred due to difficulty in making a transaction and/or delay in receiving products. Consumer economics tells us that consumers always want maximized benefits with minimized cost. One primary contributing factor for the rapid growth of online shopping is convenience. People enjoy the flexibility of shopping at anytime and from anywhere with more choices and more information. Without trips to shopping malls, transportation cost is avoided.

> Without jammed traffic and long waiting lines, time is saved. Without pressure from sales people, we feel at ease. However, when slow Internet speed and disorganized Web sites significantly increase online shoppers' mental efforts and give them a frustrating experience, it is not surprising to find that online shoppers turn to "window shoppers."

Previous studies have suggested that the concerns about online shopping risks have caused people's attitude (e.g., trust) toward e-commerce to vary across countries, and is a primary factor of the acceptance and development of e-commerce in one country (Bin et al. 2003; Efendioglu and Yip 2004). Chinese online consumers perceive higher online shopping risks than their American counterparts do for to the following three reasons.

The first reason is China's low GDP per capita, which tends to increase Chinese people's financial risk, and their demands for guarantees/warranties for products/services against potential trouble and defects.

The second reason is the differences in the e-commerce infrastructural environment between the two nations. The development of the e-commerce infrastructural environment of a nation is largely determined by the nation's infrastructure investment and the level of information and communication technology and is influenced by the growth of the Internet population. The US is one of the leading nations in e-commerce because it has a well-established e-commerce infrastructural environment and the most Internet users and broadband subscribers (Liao et al. in press). In contrast, the Internet and e-commerce are relatively new and immature in China. Although many efforts have been made to improve the e-commerce infrastructure, it still lags behind the US because of the following five major issues.

1. Low Internet penetration rates: China has the second largest population of both Internet users and broadband subscribers in the world and the population of the Internet users has been predicted to pass the US to become the largest by 2009 (CNN 2007). However, both the Internet and broadband penetration rates of China are low due to the large population of China (United Nations Conference on Trade and Development [UNCTAD] 2002).
2. Inadequate network security: Chinese industries face considerable network security problems, which partly explain why online payment is not favored in China. The security issues are mainly because security has lagged behind systems development. For some industries, information systems were developed entirely without security safeguards or with network equipment that has not been tested. Additionally, computer networks are vulnerable to vicious hackers because of the usage of open operating systems with low security (UNCTAD 2001). As a consequence, many Chinese enterprises have to disconnect their intranets from the Internet to guarantee network security.
3. Lack of credit cards and a nationwide credit card system: China's e-commerce is challenged by its cash-oriented economy (Internet World Stats 2004). Most Chinese commercial banks only issue debit cards and credit cards are owned by a quite small portion of the Chinese people. According

to market research conducted by the McKinsey (China) Ltd. (Zhang 2005), among China's population of 1.3 billion, only approximately 7 million Chinese people have a credit card. Moreover, the 2% credit card household penetration rate in China is quite low compared with more than 75% in the US (McKinsey [China] Global Institute 2006). The popularity of credit cards as a financial instrument in China has been hindered by the absence of a single nationwide electronic payments clearing system: China's commercial banks and other financial institutions are not fully electronically interlinked yet. Bank cards are issued by commercial banks independently, and bank card payments can only go through the card issuer. To process electronic payments by different bank cards, an online retailer must connect to many fragmented payment systems, which significantly increases the costs for both retailers and consumers (UNCTAD 2001). As a result, most e-commerce companies in China stick with offline payment methods for online transactions.

4. Inefficient logistical systems: Logistical infrastructure foot-dragging, such as insufficient transport networks and an inefficient national postal system, is the main challenge to fast and prompt delivery. To work around the logistical difficulties, most e-commerce companies in China either work with specialized delivery firms or build up their own delivery teams and distribution networks to meet their distribution needs (UNCTAD 2001).
5. Lack of e-commerce-related laws and regulations: It is common worldwide that laws lag behind in cyber society. The law makers in China have a harder time keeping pace with the complex issues arising from the use of the Internet and e-commerce due to their relatively short existence and rapid development. "There has not been a comprehensive law addressing all aspects of e-commerce to date. The rapid growth of Internet applications has found many government institutions without a mandate to address many of the legal issues arising from the emergence of e-commerce" (UNCTAD 2001, 245). As a result, Chinese online consumers sometimes find that there are few legal actions they can rely on to protect their rights. Additionally, under the influence of Confucianism, the business system in China operates by laws and obligations as well as personal relationships. In some cases, contracts can be changed and promises may even be broken (Zhang et al. 2002). Consequently, Chinese online consumers tend to be skeptical of e-commerce companies' policies and discredit the companies' promises. Although, significant progress has been made by issuing laws, regulations, and rules to accommodate e-commerce in recent years, more efforts are still needed to create a legal and regulatory environment that defines standard norms of e-commerce practice and restore credibility of e-commerce companies.

Based on this discussion, it can be concluded that China's relatively immature e-commerce infrastructure makes Chinese people feel threatened by high risk and prevents them from converting to online shoppers. They are likely to suspect the reputations of online retailers and are concerned about product warranties, postpurchase assistance, and how to return a purchased product if they are not satisfied with it, because they sometimes may find that there is no place to go to settle e-commerce disputes, and it is time- and energy-consuming to return a product due to the immature logistical systems (Bin et al. 2003).

The third reason is the social and cultural differences between China and the US. Human beings are social animals and our first instinct is to trust others. As individualists, the default trusting stance of American people is to trust all others until there is a reason not to. By contrast, as collectivists, Chinese people are sensitive to the ingroup-outgroup boundary. It is more difficult for them to establish mutual trust with outgroup people than with ingroup people. The differences in trusting stance toward outgroup members make Chinese people more risk-averse and perceive higher levels of risks than American people do (Jarvenpaa et al. 1999).

The gap in the trusting stances between the two cultures is widened in the virtual online world. Chinese culture prefers high-context communication patterns, and Chinese people tend to form their opinions of someone's trustworthiness based on other people's agreeableness and their relationships with others (Efendioglu and Yip 2004; McCort and Malhotra 1993). Therefore, face-to-face socialization and in-person connection is very important for Chinese people to create closeness and provide emotional support. Indeed, the Internet opens a new interactive approach to socialization; however, it is less interactive than face-to-face communication in that large amounts of important contextual information cannot be conveyed effectively and efficiently through faceless online communication (Sproull and Kiesler 1991). Although the negative impact of online communication on individuals, organizations, communities, and society exists in both cultures, it causes a relatively more profound consequence in the Chinese culture. For example, the anonymity within the online community makes Chinese Internet users feel more threatened by and more vulnerable to Internet predators and Web identity thieves, and thus leads to a negative effect on Chinese online shoppers' trust of e-commerce. Moreover, under the widespread belief that seeing is believing, most Chinese consumers may be more uncomfortable than their American counterparts without seeing and touching a product prior to purchase, especially given the existence on China's market of an abundance of counterfeit and low-quality products, some of which may be harmful to consumers' health.

Uncertainty avoidance has been claimed to be another source of perceived online shopping risks. For example, Japanese consumers have fairly exacting demands for product safety and guarantees against trouble, accidents, and defects and tend to focus on economic and credibility information sources, as they have high uncertainty avoidance (Madden et al. 1986; Okazaki 2004; Okazaki and Rivas 2002). It may seem paradoxical that Chinese consumers perceive higher levels of online shopping risks than their American counterparts do, yet China, according to Hofstede's model of culture, has lower uncertainty avoidance than the US (see Chapter 2). There are two reasons for this paradox. The first reason is that uncertainty avoidance is not equivalent to risk avoidance. Risk is a relatively more specific and concrete concept and "often expressed in a percentage of probability that a particular event may happen" (Hofstede 2001, 148). Uncertainty is a more diffuse feeling about the unknown aspects of the future and has no probability attached to it (Hofstede 2001). The second reason is that given that the difference between the uncertainty avoidance levels of China and the US is not large, the effects of social, cultural, and e-commerce infrastructural characteristics of China mentioned earlier are large enough to cancel out and even exceed the effects of the low level of uncertainty avoidance of China.

It should be noted that contradictory findings exist regarding the level of uncertainty avoidance of China. For instance, Ji and McNeal (2001) claimed that China was a strong uncertainty avoidance culture. In addition, China's low uncertainty avoidance score might seem questionable, considering that both Japan and South Korea are geographically adjacent to, and have the same cultural origin (ancient China), as China, but they score much higher than China on uncertainty avoidance (92 for Japan, 85 for South Korea) (Hofstede 2001). In the future, more research effort is needed to address the relationship between the levels of uncertainty avoidance and perceived purchasing risks of Chinese online consumers.

4.2.2.2 Purchase Advice

Purchase advice is an important component of the function factor in Table 3.15. It was speculated that Chinese consumers would perceive purchase advice as more important than American consumers would due to the cultural differences on power distance (see discussion in Section 2.3.2.1), which influences the way people accept and give authority. In high power-distance cultures, people show a strong respect for and dependence on authority and are prone to accept authority without question. In contrast, low power-distance cultures encourage people to argue, express disagreement with, and even criticize authority. Compared with the US, China is a relatively higher power-distance culture. Because research institutions/organizations possess more knowledge, information, and research capabilities empowered by advanced research facilities than normal consumers do, Chinese consumers are more likely than American consumers to perceive the research institutions/organizations as powerful authorities and consider purchasing advice from those institutions/organizations as more credible. However, it is interesting to note that African-American participants scored significantly higher than Caucasian-American participants did on all items of purchase advice (see Table 4.1), but the difference between American and Chinese participants was not significant.

5 Guidelines for Content Preparation

5.1 OVERVIEW

To design a Web site, usable or unusable, is hard work. Nobody wants to waste their time, energy, and money on designing an unusable Web site. But how does a designer design a usable one? What differentiates usable Web sites from unusable ones? The answer is user experience. There are many design guidelines on the details that designers need to take into account to provide an enhanced user experience, unfortunately, most of them tend to focus on secondary issues, such as aesthetics and navigation (Norman 1988), and neglect the most fundamental component: content, which is the prerequisite for effective Web site operation. User experience is not just about looking good. Without the correct content, the best aesthetics and navigation in the world will do us no good. Although some studies offered design guidelines regarding the information content to include on e-commerce Web sites (e.g., Fang and Salvendy 2003; IBM 2004; Rohn 1998), the benefits of these guidelines are limited because they only cover part of the information content on e-commerce Web sites, and some of them are quite general.

Then, how does one design a Web site with high content usability? Some designers prepare the content based on their intuition. Some others think that the more information they put on their Web sites, the better for the users of their sites. Neither approach is correct because they leave the end users of their products out of the design process and ignore the variability in their information needs, which vary with their ages, genders, education, occupations, culture backgrounds, and such. One effective solution for high content usability is user-centered design, which can be defined as a design philosophy and a process focusing on the needs and wants of end users at each stage of the development process. Under this design approach, designers are required to prepare Web site content from the perspective of end users and based on their understanding of the audience of their sites.

It should be noted that this discussion about the impact of content usability on user experience applies to information appliances as well. For example, according to Guo and Salvendy (in press), user satisfaction with cell phones was influenced by the quality of the cell phone content, and users indicated that the content was not sufficient.

The general information factor structure in Table 3.15 can be used as a checklist for both preparing content information for and evaluating content usability of Web sites and information appliances (see Tables 5.1 and 5.2 for checklists for evaluating the content usability of e-commerce Web sites and information appliances, respectively). In the following sections, based on the discussion in previous chapters, design guidelines are derived to improve content usability (see Table 5.3 for a summary of design

TABLE 5.1
Content Usability Evaluation Checklist for E-Commerce Web Sites

Content Usability Evaluation Checklist for E-Commerce Web Sites
Does the Web site provide information on product economic attributes?
Does the Web site provide information on product sensory attributes?
Does the Web site provide information on product performance, quality, and value-retention capability?
Does the Web site provide information on the new technology used in products?
Does the Web site provide information on product composition?
Does the Web site provide information on product convenience features?
Does the Web site provide information on the manufacturer of a product?
Does the Web site provide information on product warranty, guarantee, and postsale assistance?
Does the Web site provide information on transaction security?
Does the Web site provide information on privacy policy?
Does the Web site provide information on return and exchange policies?
Does the Web site provide information on product shipping policy?
Does the Web site provide updated information on shipment status?
Does the Web site provide information on how to contact a human representative?
Does the Web site provide convenient search functions in terms of product attributes, prices, etc.?
Does the Web site provide purchasing advice?
Does the Web site provide comments, reviews, and testimonies from customers or other third-party institutions?
Does the Web site provide an operational guide, aid, and support for online transactions (e.g., FAQ, shopping cart)?
Does the Web site provide the compare function to help customer compare products?
Does the Web site allow its customers to establish and manage their own accounts and maintain necessary records?

TABLE 5.2
Content Usability Evaluation Checklist for Information Appliances

Content Usability Evaluation Checklist for Information Appliances
Does the information appliance provide a search function?
Does the information appliance provide information on all its functions (e.g., multimedia functions)?
Does the information appliance provide help information?
Does the information appliance provide information on current status?
Does the information appliance provide information on how to return to previous status?
Does the information appliance provide information on how to input commands to the information appliance?
Does the information appliance provide information on its manufacturer?
Does the information appliance provide information on its service provider?
Does the information appliance provide feedback information to its users' actions?

TABLE 5.3
Summary of Design Guidelines

Information Content Items	Design Guidelines
Security and privacy content	Security and privacy policies should be presented in a conspicuous place of every webpage, updated regularly, and explained without ambiguity.
Performance and new technology	Product performance and new technology information should be presented in detail. In addition, e-commerce Web sites should allow their users to compare products in terms of product performance and new technology.
Warranties and postsale assistance	E-retailers should provide detailed information on product warranties and postsale assistance. In addition, they should allow their customers to compare products in terms of these information items.
Product return and exchange	Policies on product return and exchange should be presented in a conspicuous place of every webpage, updated regularly, and explained without ambiguity.
Shipping	Shipping policies should be presented in a conspicuous place of every webpage, updated regularly, and explained without ambiguity. E-retailers should provide updated shipment status information.
Service and functions	E-commerce Web sites and information appliances should provide detailed information on all their services and functions.
Sensory attributes	Product sensory attributes should be presented in detail and without ambiguity. Pictures from different angles, multimedia technology, and free sampling are strongly encouraged to provide information on product sensory attributes.
Contact information	Contact information of e-retailers, such as telephone number, mailing address, e-mail address, and fax number, should be easily accessible.
Search function	Both Web sites and information appliances should provide search functions in terms of various attributes, such as price, date, and brand.
Manufacturer	For Chinese online shoppers, manufacturer information is strongly recommended to make a product stand out from those of other competitors.
Purchasing advice	E-retailers should provide information to help their customers find a satisfying product. Comments, reviews, and testimonies on products from customers or other third-party institutions are strongly recommended.
Economic attributes	Product economic attributes, such as price and discount, should be presented in a conspicuous place. E-commerce Web sites should also allow their users to compare products in terms of economic attributes.

guidelines) with a focus on e-commerce for American and Chinese e-shoppers and information appliances. The primary goal of these guidelines is to improve content usability by providing information that meets end users' needs. Because the perception of attractiveness and the appeal of a product can be augmented by the information conveyed to consumers, the guidelines can also provide general rules and suggestions for e-commerce retailers and marketers and information appliance manufacturers to enhance their products' marketability and purchasability by incorporating or emphasizing the information that is preferred by users in the advertisements for their products. Moreover, the provided guidelines highlight the major differences in information preferences between Chinese and US online consumers; therefore, the guidelines can help international e-commerce companies who are targeting online consumers in China or the US, especially US international e-commerce companies, adjust and localize their marketing strategies to make their products more appealing and attractive to online consumers in China or the US.

The checklists and guidelines can be used throughout the development process of a product. In the very early stage, the checklists and guidelines can be used to analyze the requirements of users for the content. In the middle development stage, they can provide developers with insight about how the content should be presented. For example, Web site designers may want to post in a conspicuous place the information that online shoppers expect. After a product is designed, the checklists and guidelines can be used to help evaluate its content usability and provide suggestions for improvement.

Given the various influences on content usability, such as Web site types, product categories, and user characteristics, it is sometimes impossible to provide concrete recommendations and solutions to problems. What is needed instead is a general knowledge of the influencing factors and the underlying mechanisms of how those factors influence content usability in order to choose the best technical solution in a particular situation. This results in certain guidelines in this chapter being of a more descriptive nature. It should be noted that because the guidelines were developed from the findings based on a limited number of product categories and highly homogeneous sample populations, caution should be taken in applying the guidelines to other product categories and other population of users (see discussion in Section 5.3).

5.2 CONTENT PREPARATION GUIDELINES

5.2.1 GUIDELINES FOR REDUCING PERCEIVED ONLINE SHOPPING RISK

Perceived online shopping risk is a major barrier to user trust and loyalty regardless of where an e-commerce Web site originates. In Studies 1, 2, and 3 discussed in Chapter 3, online transaction security was revealed as one of the most important information items. The consistency among those studies provides evidence that online shopping risk is universal and exists in every culture. In this section, general design guidelines are first proposed on how to reduce each type of perceived shopping risk. Then, with reference to African-American, Caucasian-American, and Chinese consumers, discussion is made to illustrate how culture may affect the applications of those design guidelines.

5.2.1.1 Perceived Financial Risk

Security is an essential feature as well as one of the biggest challenges of e-commerce business. Each year, millions of dollars are spent on network security and the spending is growing at double-digit rates, yet, according to a study by London-based TNS PLC, a market research company, about 75% of online shoppers abandoned an online transaction because of security concerns in 2005 (Gold 2006). Because online shoppers prefer to shop with large and well-known e-retailers, the figure may be even higher for e-retailers with less brand awareness. E-commerce security is a complicated issue and requires both managerial efforts and technical support. Thus, the security vulnerability facts cannot be changed overnight; however, the good news is that the perceptions can be changed significantly. For example, in the TNS PLC study referenced above, among those 75% of online shoppers, 90% indicated that they would have placed an order if they had seen a recognized security marker (Gold 2006). The following are several guidelines for building customer confidence in the security of your e-commerce Web sites.

- The key to effective e-commerce security is a good security policy. The policy should be put in writing, endorsed by senior management, and available on your Web sites. In the policy, how security measures will be carried out and enforced should be explained clearly, and the policy should be continually updated to keep pace with technological advances.
- Explain how antivirus technology and firewalls are employed to ensure a virus-free transaction environment, and how advanced encryption techniques are used to ensure the confidentiality and integrity of sensitive transaction information (e.g., credit card numbers) transmitted on public and private networks.
- Place a security logo, such as those of GeoTrust, Verisign, and Hacker Safe, on all webpages and at all points in checkout. Consumers look for the seal of approval, especially tech-savvy and cautious online shoppers, who are quite responsive to the appearance of security trust marks. One study by IBM in 2006 reported that 70% of consumers would only shop at Web sites with a recognized security protection seal (Gold 2006). Additionally, it is interesting to note that a study by Petco, a pet supply retailer, indicated that the effects of security logos varied with their locations, with the location on the upper left between the search box and the navigation bar being most effective and the location below the footer on the lower right being least effective (Gold 2006). Although the best locations for security logos for a specific Web site may depend on the characteristics of that site, it is recommended to place security logos along visitors' eye flow path, such as the top location underneath the Web site's logo. It should be noted that the effects of security logos are greater for less-known e-commerce Web sites, because consumers do not know if these Web sites are legitimate businesses and thus are more reluctant to shop with them. It has been reported that by the company that sells the Hacker Safe product that while most well-known e-commerce Web sites experience single-digit sales increases, less-recognized companies experience double-digit increases, in some cases as much as 30% (Gold 2006).

- Provide online shoppers with help to set up strong passwords for their accounts. In this way, online shoppers, especially those who are less tech savvy, feel more confident by knowing how well they are protected and that they have the control over the protection level. This is also a good strategy for e-retailers to forge a bond with their customers and gain their trust by showing that they care about their customers and are willing to help at any time.
- Add testimonials and third-party endorsements such as recognized magazines, industry experts, and associations to boost the reputation of e-retailers. Trust has a transitive property. Online shoppers' trust of information sources and organizations with high credibility can be transmitted to e-retailers by quoting positive comments from those sources and organizations.

5.2.1.2 Perceived Performance Risk

Perceived performance risk is mainly caused by the inabilities of consumers to inspect a product due to the lack of tangibility in transacting over the Web. By and large, products can be divided into the following two categories:

1. *Tangible products* are products that have physical dimensions and attributes discernible by senses. For example, clothing and consumer electronic products are tangible products.
2. *Intangible products* are products that cannot be touched and do not have observable characteristics. Services, music, and computer software are examples of intangible products.

In brick-and-mortar stores, consumption experience of tangible products and some types of intangible products can be easily obtained through free trials. For instance, we can try on clothes, taste food, work the controls of digital cameras, and listen to music. This is quite an effective marketing strategy in that product experience assures consumers about the value of what is being considered, especially for hedonic and luxury products, the consumption of which involves multisensory, fantasy, and motive aspects. Such strategy has been employed by some online marketers to deliver product experience to their customers and has significantly reduced product returns. For example, we can listen to trial music at www.emusic.com, browse free book chapters at books.google.com, download software that is free in a limited period of time at www.adobe.com, and get free quotes on car insurance at www.geico.com. However, the products that can be sampled online are limited and mostly fall in the intangible product category. The inability to deliver product experience of tangible products results in high volume product returns. For example, the return percentage of clothing is high, around 27%. What can we do to reduce consumers' skepticism over the performance of other products? The following are several suggestions.

- Visual presentation: "A picture is worth a thousand words." This is particularly true on the Internet. People do not read text on webpages, rather they scan. That means that paragraphs of words about product performance are likely to be ignored. Thus, for tangible products, product pictures or animations can quickly capture consumers' attention and considerably compensate for their inability to touch the products before purchase, especially when pictures or animations can reveal the products from different angles. For intangible products, logos are recommended to make the products more appealing by adding an element of tangibility to the products. For example, the raging bull logo of Merrill Lynch (www.ml.com) vividly illustrates their services. It should be noted that the use of visual presentations should not degrade Web site performance, such as long loading time, which will give rise to a frustrating shopping experience.

- Samples: For products that cannot be sampled online, free samples can be mailed to consumers upon their requests. For expensive products, a cheaper version of the product for sampling purposes can be offered at a low price.
- Testimonials and reviews: Consumer testimonials and reviews with full names on product quality and postsales services not only can serve as advertisements for the products, but also can serve for the purpose of product experience sharing and answering questions of potential buyers.
- Online assistance: Unlike in real stores, where consumers can talk face to face with sales people regarding questions and concerns about products, online shoppers often find themselves in a situation where the information provided online cannot adequately address their questions and concerns, or they do not know where to seek help and suggestions. Online assistance, such as online chatting and a hotline, is an effective tool to build a good relationship with customers and win business.

5.2.1.3 Perceived Psychological Risk

Privacy concerns play a role as important and prevalent as security issues in affecting corporate reputations and brands of Internet companies and hindering the adoption of the Web by the online customer to the next level. However, privacy concerns have received less attention than security issues have. In some countries, such as the US, it is not a legal obligation to post privacy policies on the Web, but it should be noted that an increasing number of sites in those countries have chosen to adopt privacy policies despite the increased legal risk. Although this fact reflects the growing public awareness of the importance of online privacy, the efforts of the whole Internet community to protect customer privacy is still not enough. Generally, larger companies are more respectful of customer privacy than smaller companies in the same industry are, and travel and airlines industries tend to provide more privacy protection than pharmaceutical and health care industries (Mello 2005).

Typically, privacy information is collected for two main marketing purposes. First, some information, such as names, physical and e-mail addresses and other personal data, is necessary to complete a transaction and provided voluntarily by online shoppers. Second, some information is passively collected by e-commerce sites to better understand and serve their customers. For example, the technology of Internet cookies is often used to collect information about online shoppers' browsing habits and preferences, and the information is then analyzed in conjunction with customer demographic data for effective advertising and customer targeting.

The perception of privacy threats seems to vary with the characteristics of Internet users. For example, female Internet users are more likely to have higher levels of concerns with privacy invasion (Ackerman et al. 1999; Westin 1998). This may be because there are fewer female Internet users, and female Internet users on average have less Internet experience than their male counterparts do (EMarketer 2008; Pew Internet and American Life Project 2005a, 2005b). In terms of their attitudes toward privacy, online customers can be divided into three groups (Ackerman et al. 1999).

- *Privacy fundamentalists* are extremely concerned about any use of their personal information and generally unwilling to provide their personal information on the Internet even when privacy protection measures are in place. Privacy fundamentalists constitute a relatively small group of Internet users.
- *Privacy pragmatists* are less concerned about privacy threats than privacy fundamentalists, and their concerns can be significantly reduced by the presence of privacy protection measures. More than half of the Internet users can be classified as privacy pragmatists.
- *Marginally concerned users* are willing to provide personal data on the Internet under almost any condition. The number of marginally concerned users falls between those of privacy fundamentalists and pragmatists.

Given the different levels of privacy concerns among Internet users, a one-size-fits-all approach to online privacy is unlikely to succeed. Internet companies need to enact their own privacy policies based on analysis of the audience of their sites. Because privacy pragmatists are the largest Internet user group, it is logical to assume that they should be the target of most Internet companies as the first step to decrease privacy concerns. The following guidelines are suggested to reduce privacy concerns.

Although it is reported that the existence of privacy policies, privacy seals, and personal data retention policies does not significantly reduce Internet users' privacy concerns (Ackerman et al. 1999), we still strongly encourage Internet companies to post them on their sites. For those companies that are not legally obligated to post privacy policies, the existence of privacy policies, privacy seals, and personal data retention policies can make them stand out from their competitors by improving their public images and trustworthiness.

Their privacy policies should be easy to understand and focus on addressing the following six major privacy concerns (Ackerman et al. 1999; GVU 1998):

1. Is the personal information of Internet users going to be shared with other companies and organizations? If yes, with whom?
2. What information is going to be collected? What are the purposes for collecting such information?
3. How is the information going to be used? Is the information going to be used for marketing without permission? How can users control user their data for marketing?

4. Is the information going to be used in an identifiable way or in an aggregate form?
5. Is the information going to be stored in Internet companies' databases?
6. Can customers be easily removed from mailing lists and can their personal information be destroyed upon request?

Among the various types of personal information, Internet users seem to be most concerned about their health, income, and phone number information; therefore Internet companies, when possible, should avoid forcing their customers to provide such information. Compared with e-mails and postal mail, unsolicited phone calls are more likely to be perceived as privacy violations because they can give rise to a higher level of annoyance, especially when it is difficult to end the annoyance (Culnan 1993; Westin 1991). As such, when contact information is needed, Internet companies should ask Internet users to provide their e-mail and/or postal addresses and make phone numbers optional. When phone numbers are necessary, companies should provide an explanation why their phone numbers are needed. It is interesting to notice that marginally concerned users are quite concerned about whether they can be easily removed from Internet companies' mailing lists upon request (Ackerman et al. 1999). Because marginally concerned users have the lowest levels of privacy concerns, Internet companies can easily win marginally concerned users' trust by simply explaining what to do to get them removed from the mailing lists.

5.2.1.4 Perceived Inconvenience Risk

Clicking the "buy" button initiates the messiest part of e-commerce, order fulfillment logistics, which accounts for around 40% of the cost of e-commerce and includes activities essential to transaction volume and amount, such as payment processing, product delivery, and possible product returns (Bayles 2001). For large companies, order fulfillment is usually done in a completely automatic and seamless manner; however for small companies, the process often relies relatively more on less automated traditional methods, such as human labor, and causes more errors and delays, which in turn initiate a sequence of other problems to significantly degrade customer satisfaction and loyalty and even turn a profitable sale into a debt. In addition to logistical challenges imposed by their own logistics capability and management, e-commerce companies also face other various barriers imposed by the economy and society they are in, such as logistics infrastructure of the economy and the political, legal, and cultural environments of the society (see Section 4.2.2). This makes e-commerce logistics a complicated issue and calls for the efforts of individual e-commerce companies, as well as the whole logistics industry and society. For individual e-commerce retailers, the following guidelines are suggested to improve their order fulfillment process.

E-commerce retailers should provide estimated shipping times. Patience is one of the prices online shoppers need to pay for online shopping convenience. However, the timing of delivery is important to customers under some circumstances. For example, when we buy a birthday present online, we normally want the present delivered on or before the birthday. Thus, accurately estimated shipping time can allow customers to make a better informed decision by lowering their uncertainties. It is more important to keep customers informed of the shipping time during shopping seasons, such as for Christmas, because longer shipping time is often expected due to the overloaded logistics infrastructure.

The companies should provide estimated shipping costs. Although online shoppers are not responsible for moving the goods from the warehouses to their destinations, they still bear the costs for it. Different shipping costs may be charged depending on (1) the size and weight of the goods, (2) how soon the goods need to be delivered, and (3) the distance between warehouses and destinations. Thus, as with shipping time, clear policies on shipping cost rates and accurate shipping cost estimates can enable customers to make well-informed decisions about whether to purchase a product online and how they would like the product delivered, especially when the shipping costs are considerably larger compared with product prices.

The retailers should provide shipment status tracking. In addition to providing up-to-date information on order status, shipment status tracking, provided by many third-party logistics companies and distributors, offers customers an opportunity to get involved in the order fulfillment process, which can be very helpful in reducing customers' uncertainties caused by prolonged shipping time, such as with international shipping. Furthermore, by allowing customers to track their goods online, e-commerce companies can significantly decrease incoming telephone call inquiries on order status.

E-commerce retailers should handle product returns well. Product returns are inevitable due to damages, defects, customer dissatisfaction, recalls, and other reasons. Thus, in addition to sending goods to customers, e-retailers are also required to have the capability to send goods back to themselves, which is usually referred to as reverse logistics. Indeed, product returns cause significant economic losses up to billions of US dollars each year. However, retailers continue offering liberal and gracious return policies and sometimes even tolerate customers abusing their policies, because they view product returns as a powerful competitive weapon to build customer relationships and increase customer satisfaction, which is the most important strategy to maintaining a competitive advantage and winning market share. Compared with brick-and-mortar stores, e-retailers are often faced with much higher product return percentages, particularly for tangible goods such as clothing, and the volume of product returns is skyrocketing (Bayles 2001). This is primarily because of the inability to perform adequate performance evaluation before purchase (see Section 5.2.1.2). Furthermore, competitors are just a click away in the online world. If product returns are not handled smoothly, customers will immediately leave a company's site and will never come back regardless of how good that company is in other aspects. As a result, online merchants are more pressured by reverse logistics challenges than their offline counterparts are. The challenges cannot be tackled overnight. First, e-retailers must recognize the importance of reverse logistics and then set it a business priority and create policies to enforce that priority. Posting clear product return policies on the Internet is an effective strategy to achieve a successful product return process. Unlike what was suggested by the experiments by Liao et al. (in press), most online customers want to know how a product can be returned before they make a purchasing decision (Bayles 2001). The return policies should clearly address the following issues that online shoppers are mainly concerned about:

- Whether they can get a refund or exchange after a product return.
- Whether they will be charged for product returns.
- The time period in which a product return can be made.
- Detailed return instructions, such as how to request a return authorization, where to send the return, and how to contact the shipper.

The characteristics of audience vary across the Web sites. As mentioned earlier, under the user-centered design principle, these design guidelines suggested above should be used based on the analysis of the characteristics of users of each site. Generally speaking, e-commerce Web site users can be divided into the following three groups based on their shopping experiences (Forsythe and Shi 2003):

1. *Browsers* are users who browse e-commerce Web sites but do not place an order when they have decided to purchase a product. They tend to be 11 to 20 years old and have less than 1 year of Internet experience and low incomes. This makes sense because people at ages of 11 to 20 may not have credit cards and well-paid jobs, so they are unable to purchase online.
2. *Heavy shoppers* are users who place an online order all or most of the time when they have decided to purchase a product. Compared with browsers, heavy shoppers are more likely to be older male users in ages of over 50 years with more than 4 years of Internet experience and higher levels of incomes.
3. *Moderate shoppers* are users who place an online order half of the time or less often when they have decided to purchase a product. The profiles of moderate shoppers tend to fall between those of browsers and heavy shoppers.

The perception of the online shopping risk of these three groups is a function of their demographic profiles. Among the four types of perceived online shopping risks, browsers tend to perceive more financial risks, time/convenience risks, and psychological risks than moderate and heavy shoppers do, whereas heavy shoppers tend to perceive lower levels of risks in all risk categories than browsers and moderate shoppers do. Moderate shoppers are more likely to perceive product performance risk than the other two groups would. This discussion implies that although reducing perceived online shopping risk is essential to the success of every e-commerce Web site, e-commerce Web site designers can build user trust and loyalty more effectively by understanding their core markets and Web site users and by identifying the major sources of perceived risk. For example, although women are catching up, women, particularly adult married ones, tend to lag behind men in the online world. One reason for the disparity is that women tend to be relatively less tech savvy than men are. As a result, compared with their male counterparts, female Internet users tend to be more concerned about risks associated with Internet use and tend to be browsers and moderate shoppers. Similarly, because Chinese online consumers are relatively young and inexperienced with the Internet, most of them fall in the categories of browsers and moderate shoppers. The same logic applies to African Americans, because they tend to have less Internet experience in comparison with Caucasian Americans. One implication of this discussion is that for e-merchants selling merchandise mainly for female consumers, African-American consumers, and/or Chinese consumers, they face more challenges to turn these users into heavy shoppers. According to Gefen (2000), familiarity builds trust in the online world. To gain the trust of browsers and moderate shoppers, e-marketers can make their customers and potential customers more comfortable and familiar with e-commerce–related technology by providing the necessary technical assistance on their Web sites. For example, considering the fact that credit card penetration rates are relatively low in China, e-retailers can reduce the perceived financial risk of Chinese online shoppers related to credit card use by emphasizing their sites' secure and private transaction environment and explaining the technology, such as data encryption, used to protect the shoppers' credit card information.

E-retailers in China suffer to a greater extent from reverse logistics issues than those in the US, given that the logistics infrastructure in China is less mature, that there is less legal protection in China for Chinese consumers, and that Chinese consumers are less likely to believe that e-retailers will enforce their promises and policies. Therefore, posting product return policies and detailed product return instructions are extremely important for e-retailers in China. Additionally, they should stress the convenience of their product return process.

5.2.2 Guidelines on Product Attributes and Functions

For African-American and Chinese online consumers, e-retailers should give priorities to emphasizing utilitarian and economic aspects of a product by stressing the product's performance and cost-effectiveness. With the increase of the prices, the emphasis should be tilted to the economic aspects of the products. It should be noted that it is wrong to simply assume that Chinese online consumers are less willing than American online consumers to buy high-priced products because Chinese online consumers are more likely to base their decision on cost-effectiveness and weigh marginal benefit against marginal cost. Thus e-retailers in China can increase their sales, especially those of high-priced products, by convincing Chinese consumers of the benefit-cost potential of the products.

For information appliances, information about their functions is important. Guo and Salvendy (in press) reported that the information could be divided into two categories: information for specific functions and information for general functions and operations. Users' information needs for these two categories vary with factors such as age and education level. Take cell phones as an example. Younger users, especially students, are attracted to multimedia functions, but more elderly users or industrial populations do not consider them as important (Guo and Salvendy in press). As a matter of fact, the Jitterbug cell phones (http://www.Jitterbug.com), which do not have many fancy features but are simple to use, have been designed specifically for senior citizens. As discussed in Study 4 in Chapter 3, another example is the text input function of cell phones. Information about this specific function is essentially important for Chinese users, because text messaging is very popular in China but the Chinese language, compared with alphabetic languages, is relatively more difficult to input on a small cell phone panel. The implication for information appliance manufacturers is what we have repeated several times so far in this chapter: to prepare content based on the characteristics of targeted users.

Making product comparisons in terms of product attributes and functions easy, such as tabulating product information, will facilitate online shoppers' purchasing decision-making processes and enhance their shopping experiences. Although country information is not as important to Chinese consumers as other items are, given that the differences between the two cultural groups in this item were highly significant, for Chinese online consumers, e-commerce Web site designers can make a product stand out from others by stressing the country information. Safety features and functions are another selling point to Chinese consumers. For example, the radiation levels of cell phones and computer monitors are major concerns for Chinese consumers.

For American online consumers, in addition to product performance, e-retailers and information manufacturers should give priorities to emphasizing hedonic aspects of a product by stressing the product's performance, convenience features, and any new technology utilized in the product. Making product comparisons in terms of the aforementioned information items easy—for example, by presenting products in a table format—will facilitate American online consumers' purchasing decision-making processes and enhance their shopping experiences. Furthermore, compared with Chinese online consumers, e-retailers and information appliance manufacturers may want to provide enough information about product composition and accessories.

5.2.3 Limitations and Recommendations

Like every other study, the studies presented in this book have their own limitations due to their limited scope and other restrictions. First, although the participants in each study were considered to be a representative sample of the population of interest, the high homogeneity of the sample populations may limit the generalization of the study results. For example, the participants in the studies of Liao et al. (2008, 2009, in press) were American and Chinese college students. They were considered to be a representative population of Internet users because most Internet users in China and the US are young generations with a college and/or postcollege degree. However, whether the study findings can be generalized to other groups of online consumers is questionable for two reasons. First, as the Internet penetrates every corner of the world, the Internet is no longer the privilege of young generations and the composition pattern of Internet users is changing over the years. Second, college students may be heavy Internet users, but they have relatively limited consumption power due to their income levels. As mentioned earlier, heavy online shoppers tend to be relatively affluent and senior people in the US, which implies that the college student sample may deviate to some extent from a nationally representative sample. Therefore, further validation is needed before the findings presented in this book are applied to the whole spectrum of online consumers.

Second, the discussion in this book mainly focused on e-commerce, therefore the implications of the study findings and the utility of the guidelines may be limited for other types of Web sites. Although e-commerce was chosen because it is ranked among the top two of all domains for visitation and use, it is not our intention to indicate that content usability is not important for Web sites with low ranking in terms of visitation and use. As indicated in earlier chapters, the content of a Web site is the foundation of that site and directly determines site operation and user experience. As such, Web site designers should set content preparation as a priority from the beginning of their Web site development process.

Third, the comprehensive information factor structure proposed in Table 3.15 has three limitations. First, although it is our intention to encompass all the information items for e-commerce Web sites, it is not practical to include all the information, especially Web site and/or product-specific information, because there are too many types of e-commerce Web sites and too many types of products. Second, for purposes of practicality and sufficient generalizations, the information factor structure was developed at some level of abstraction. Third, the relative importance of each factor in the factor structure is not completely clear. Thus for Web site designers, for a particular type of e-commerce Web site and/or a particular type of product, the utility of the factor structure may be limited because it is not complete and detailed enough. However, we believe that the factor structure can help designers start on the right foot. In the future, more research can be done to enrich the current factor structure with more site- and/or product-specific information items. On the other hand, the absence of a comprehensive and definitive classification of information content is a major challenge in the body of research in content preparation. The current factor structure can serve as a theoretical basis for the avenue of research on a generic information factor structure encompassing all types of Web sites and/or all types of information systems. Additionally, further studies can be conducted to reveal the relative importance of the factors in the factor structure, which can provide valuable guidance with Web site designers on how to effectively allocate their limited resource for content preparation.

Fourth, the studies presented in this book were restricted to portable electronic products and information appliances. Although the product categories chosen in the studies (e.g., particularly MP3 players, digital cameras, and laptop computers) are representative of the products sold online, it is not clear whether the findings can be generalized to other consumer product categories. The experiments by Liao et al. (in press) have suggested that consumers' preferences for some information items may be influenced by product prices, and the theoretical and empirical evidence from studies in traditional marketing have indicated that consumers' information search behavior differs significantly across product categories (e.g., Bloch and Richins 1983; Girard et al. 2002; Klein 1998; Norton and Norton 1988). Moreover, as implied by the discussion earlier, online consumers' information needs are a function of products' tangibility. For example, e-shoppers may need more information to evaluate the performance of tangible products because they are more likely to perceive greater performance risk with tangible products than intangible products. Validation is needed before the study findings are expanded to other product categories.

Fifth, one of the purposes of this book is to examine cultural effects on content preparation. African-American, Caucasian-American, and Chinese cultures were chosen because American and Chinese cultures are typical Eastern and Western cultures, respectively, and African-American and Caucasian-American cultures are typical subcultures within the American culture. These cultures were compared along the cultural dimensions of Hoftede's Culture Model, cognitive styles, economic standing, and e-commerce infrastructure. It is not clear whether and how other factors, such as politics, will influence content preparation. Moreover, although the study findings presented in this book can reveal the mechanism of how cultures influence content preparation, further investigation is needed to answer the questions of whether the guidelines can be applied to other cultures and to what extent they can be applied. In the future, more research is needed to explore other potential factors and expand the study findings to other cultures.

Appendix A: Content Usability Survey for Study 1

INSTRUCTIONS

The purpose of this survey is to assess the relative importance of each item presented in the questionnaire below. In answering, please consider the following.

Imagine yourself searching the Internet for information. You are searching for information that falls into one category. Select one item below.

- Personal interests: reading the news, learning, hobbies, maps, weather, etc.
- Work-related: education, research information, problems' solutions, stocks, etc.
- Entertainment: playing online games, listening/downloading music, watching movies, etc.
- E-commerce: purchasing products/services, banking, reserving airline tickets, etc.
- Socialization and communication: chat room, online forums, joining a group, etc.

Think of the "particular" information that would be important to you in that scenario. Then, answer all of the following questions accordingly.

Read each statement and decide how strongly you agree or disagree. Then, select the most appropriate response (i.e., N/A, 1, 2, 3, 4, 5, 6, 7). Next, decide how important that information would be to you in your search (i.e., 1, 2, 3). Not Applicable—Select N/A if you feel that the statement does *not* apply (e.g., does not make sense) to your information search. Then, continue to the next statement.

Please answer *all* of the questions carefully. Your participation is greatly appreciated!

	Questions	Not Applicable	Strongly Disagree	Disagree	Slightly Disagree	Neutral	Slightly Agree	Agree	Strongly Agree	Not Important	Neutral	Important
Site												
1	The Web site provided up-to-date information	N/A	1	2	3	4	5	6	7	1	2	3
2	The revision dates of the information were extremely helpful	N/A	1	2	3	4	5	6	7	1	2	3
3	The date of creation was useful	N/A	1	2	3	4	5	6	7	1	2	3
4	I was satisfied with the currency of the information	N/A	1	2	3	4	5	6	7	1	2	3
5	The release dates of new product information was *not* valuable	N/A	1	2	3	4	5	6	7	1	2	3
Company												
6	A sufficient amount of information about the company (e.g., seller, university) was presented	N/A	1	2	3	4	5	5	7	1	2	3
7	The company's objective (e.g., mission statement) was valuable	N/A	1	2	3	4	5	6	7	1	2	3
8	Presented influential information about the retailer's reputation	N/A	1	2	3	4	5	6	7	1	2	3
9	The description of the services (i.e., gift wrap) performed by the company was thorough	N/A	1	2	3	4	5	6	7	1	2	3

Questions	Not Applicable	Strongly Disagree	Disagree	Slightly Disagree	Neutral	Slightly Agree	Agree	Strongly Agree	Not Important	Neutral	Important
10 The Web site provided the desired information of various company policies	N/A	1	2	3	4	5	6	7	1	2	3
11 The name of the company (or department) was useful	N/A	1	2	3	4	5	6	7	1	2	3
12 The company's sponsors information was *not* helpful	N/A	1	2	3	4	5	6	7	1	2	3
13 The credentials of the associated professional bodies were necessary	N/A	1	2	3	4	5	6	7	1	2	3
14 The Web site included influential press releases about the company/university	N/A	1	2	3	4	5	6	7	1	2	3
15 Sufficient information about employment opportunities was presented	N/A	1	2	3	4	5	6	7	1	2	3
Product											
16 I was completely satisfied with the product information presented	N/A	1	2	3	4	5	6	7	1	2	3
Components											
17 The information about the components of the product/services was essential	N/A	1	2	3	4	5	6	7	1	2	3

(*Continued*)

Continued

Questions	Not Applicable	Strongly Disagree	Disagree	Slightly Disagree	Neutral	Slightly Agree	Agree	Strongly Agree	Not Important	Neutral	Important
18 Great descriptions about student activities (i.e., organizations and clubs) were included	N/A	1	2	3	4	5	6	7	1	2	3
19 Product accessories descriptions were very informative	N/A	1	2	3	4	5	6	7	1	2	3
20 Components (i.e., subjects, classes) of the product/service were explained in detail	N/A	1	2	3	4	5	6	7	1	2	3
Aesthetics											
21 Description of the product appearance was comprehensive	N/A	1	2	3	4	5	6	7	1	2	3
22 The descriptions of the product's available color options was *not* useful	N/A	1	2	3	4	5	6	7	1	2	3
23 Product shapes/packaging descriptions met my needs	N/A	1	2	3	4	5	6	7	1	2	3
24 The size of the product was explained in a measurable way	N/A	1	2	3	4	5	6	7	1	2	3
Name											
25 The various names associated with the product were useful	N/A	1	2	3	4	5	6	7	1	2	3
26 The name of the manufacturer was extremely valuable	N/A	1	2	3	4	5	6	7	1	2	3
27 The brand name of the product had great influence	N/A	1	2	3	4	5	6	7	1	2	3

Questions	Not Applicable	Strongly Disagree	Disagree	Slightly Disagree	Neutral	Slightly Agree	Agree	Strongly Agree	Not Important	Neutral	Important
Durability											
28 I was pleased with the information about the product quality	N/A	1	2	3	4	5	6	7	1	2	3
29 The safety features of the product were carefully explained	N/A	1	2	3	4	5	6	7	1	2	3
30 An objective evaluation of workmanship was included in the product details	N/A	1	2	3	4	5	6	7	1	2	3
31 The information about product reliability was *not* helpful	N/A	1	2	3	4	5	6	7	1	2	3
32 The description of product/ service average lifetime was precise	N/A	1	2	3	4	5	6	7	1	2	3
33 The description of materials used in product composition was *not* useful	N/A	1	2	3	4	5	6	7	1	2	3
34 Sufficient information about the product maintenance was displayed	N/A	1	2	3	4	5	6	7	1	2	3
Product Description											
35 Product description displayed exactly what was needed	N/A	1	2	3	4	5	6	7	1	2	3

(Continued)

Continued

Questions		Not Applicable	Strongly Disagree	Disagree	Slightly Disagree	Neutral	Slightly Agree	Agree	Strongly Agree	Not Important	Neutral	Important
36	Good description of the resources provided in the facilities	N/A	1	2	3	4	5	6	7	1	2	3
37	I was satisfied with the product compatibility description	N/A	1	2	3	4	5	6	7	1	2	3
38	Sufficient information about what the product does was provided	N/A	1	2	3	4	5	6	7	1	2	3
39	Technical details of the product's performance met my needs	N/A	1	2	3	4	5	6	7	1	2	3
40	Description of the product features was useful	N/A	1	2	3	4	5	6	7	1	2	3
41	The description of product's convenience (i.e., preparation, use, or disposal) was good	N/A	1	2	3	4	5	6	7	1	2	3
42	The sensory experience (e.g., fragrance, touch, comfort) description was valuable	N/A	1	2	3	4	5	5	7	1	2	3
43	Service/product assembly information was very detailed	N/A	1	2	3	4	5	6	7	1	2	3
44	Facility description provided good information about available resources	N/A	1	2	3	4	5	6	7	1	2	3

Questions	Not Applicable	Strongly Disagree	Disagree	Slightly Disagree	Neutral	Slightly Agree	Agree	Strongly Agree	Not Important	Neutral	Important
45 The product information displayed was comprehensive	N/A	1	2	3	4	5	6	7	1	2	3
46 Excellent coverage of product nutritional content	N/A	1	2	3	4	5	6	7	1	2	3
47 The inventory information about the product was useful	N/A	1	2	3	4	5	6	7	1	2	3
48 The product description was tailored to its category (i.e., jewelry, books, electronics)	N/A	1	2	3	4	5	6	7	1	2	3
49 Deadlines needed for product availability, discounts, and/or warranties were helpful	N/A	1	2	3	4	5	6	7	1	2	3
50 Great descriptions of product limitations and terms of use	N/A	1	2	3	4	5	6	7	1	2	3
51 Totally new concept introduced in the presentation of the product	N/A	1	2	3	4	5	6	7	1	2	3
Price											
52 The price information was *not* sufficient	N/A	1	2	3	4	5	6	7	1	2	3
53 The cost of the product (e.g., fee of service/program) was precise	N/A	1	2	3	4	5	6	7	1	2	3
54 Relative pricing of the product was extremely helpful	N/A	1	2	3	4	5	6	7	1	2	3

(*Continued*)

Continued

	Questions	Not Applicable	Strongly Disagree	Disagree	Slightly Disagree	Neutral	Slightly Agree	Agree	Strongly Agree	Not Important	Neutral	Important
55	The discount (e.g., special offer, scholarship) information was useful	N/A	1	2	3	4	5	6	7	1	2	3
56	Descriptions of limited-time deals for the product/service met my needs	N/A	1	2	3	4	5	6	7	1	2	3
Reviews												
57	The purchase advice was extremely helpful	N/A	1	2	3	4	5	6	7	1	2	3
58	The results of research gathered by an “independent” research firm was influential	N/A	1	2	3	4	5	6	7	1	2	3
59	Company’s product comparisons exceeded my needs	N/A	1	2	3	4	5	5	7	1	2	3
60	Company/university’s rankings were *not* useful	N/A	1	2	3	4	5	6	7	1	2	3
61	Information of the product based on brands, prices, materials, etc., was valuable	N/A	1	2	3	4	5	6	7	1	2	3
62	Customer testimonies (dedication, brand preference, and product lifetime) were greatly influential	N/A	1	2	3	4	5	6	7	1	2	3

Questions	Not Applicable	Strongly Disagree	Disagree	Slightly Disagree	Neutral	Slightly Agree	Agree	Strongly Agree	Not Important	Neutral	Important
Membership/Account											
63 Web site provided complete details of my account information	N/A	1	2	3	4	5	6	7	1	2	3
64 My past transactions from this account was extremely useful	N/A	1	2	3	4	5	6	7	1	2	3
65 Desired information about the status of my account was displayed	N/A	1	2	3	4	5	6	7	1	2	3
66 The personalized information supplied met my needs	N/A	1	2	3	4	5	6	7	1	2	3
67 The Web site provided detailed information of my transaction history	N/A	1	2	3	4	5	6	7	1	2	3
68 The membership requirements presented was helpful	N/A	1	2	3	4	5	6	7	1	2	3
Transaction											
69 The Web site provided detailed transaction information	N/A	1	2	3	4	5	6	7	1	2	3
70 Confirmation information (e.g., reference number, receipt) was very helpful	N/A	1	2	3	4	5	6	7	1	2	3
71 The taxes applied to my purchase were explained in great detail	N/A	1	2	3	4	5	6	7	1	2	3

(*Continued*)

Continued

Questions	Not Applicable	Strongly Disagree	Disagree	Slightly Disagree	Neutral	Slightly Agree	Agree	Strongly Agree	Not Important	Neutral	Important
72 The explanation of different payment methods was *not* useful	N/A	1	2	3	4	5	6	7	1	2	3
73 The shopping cart was valuable when tracking my items while shopping	N/A	1	2	3	4	5	6	7	1	2	3
74 Information about an online application met my needs	N/A	1	2	3	4	5	6	7	1	2	3
75 The tax price was very precise	N/A	1	2	3	4	5	5	7	1	2	3
76 Presenting the amount (e.g., quantity) of items purchased was useful	N/A	1	2	3	4	5	5	7	1	2	3
Security											
77 The Web site provides a strong sense of security.	N/A	1	2	3	4	5	6	7	1	2	3
78 The security procedures implemented by the Web site was discussed	N/A	1	2	3	4	5	6	7	1	2	3
79 I felt extremely secure conducting financial transaction(s) on the Web site	N/A	1	2	3	4	5	6	7	1	2	3
80 Valuable information about security (e.g., privacy) issues was presented	N/A	1	2	3	4	5	6	7	1	2	3

Questions	Not Applicable	Strongly Disagree	Disagree	Slightly Disagree	Neutral	Slightly Agree	Agree	Strongly Agree	Not Important	Neutral	Important
Packaging/Shipping											
81 An adequate amount of shipping/delivery information was displayed	N/A	1	2	3	4	5	6	7	1	2	3
82 A sufficient amount of information about the shipment methods	N/A	1	2	3	4	5	6	7	1	2	3
83 Tracking number was great for monitoring my package movement	N/A	1	2	3	4	5	6	7	1	2	3
84 Information about the different methods of shipping was helpful	N/A	1	2	3	4	5	6	7	1	2	3
85 The expected delivery date of my package(s) was very precise	N/A	1	2	3	4	5	6	7	1	2	3
Customer Service											
86 The customer support provided by the Web site was excellent	N/A	1	2	3	4	5	6	7	1	2	3
Help											
87 The Web site understood your problems	N/A	1	2	3	4	5	6	7	1	2	3
88 Features of the Web sites were supported with great instructions	N/A	1	2	3	4	5	6	7	1	2	3

(Continued)

Continued

Questions	Not Applicable	Strongly Disagree	Disagree	Slightly Disagree	Neutral	Slightly Agree	Agree	Strongly Agree	Not Important	Neutral	Important
89 The Frequently Asked Questions (FAQ) were extremely helpful	N/A	1	2	3	4	5	6	7	1	2	3
90 Help section provided great instructions for available features	N/A	1	2	3	4	5	6	7	1	2	3
Contact Information											
91 The different forms of contact information fit my needs	N/A	1	2	3	4	5	6	7	1	2	3
92 Contact information included the desired address (i.e., e-mail and/or physical)	N/A	1	2	3	4	5	6	7	1	2	3
93 The 800 number presented was extremely helpful	N/A	1	2	3	4	5	6	7	1	2	3
94 I was pleased with the information about the physical location	N/A	1	2	3	4	5	6	7	1	2	3
Refund Policy											
95 Refund policy description was extremely comprehensive	N/A	1	2	3	4	5	6	7	1	2	3
96 Product/services warranty was tremendously thorough	N/A	1	2	3	4	5	6	7	1	2	3

Questions	Not Applicable	Strongly Disagree	Disagree	Slightly Disagree	Neutral	Slightly Agree	Agree	Strongly Agree	Not Important	Neutral	Important
97 Needed company policies were presented (e.g., cheating, late homework, security)	N/A	1	2	3	4	5	6	7	1	2	3
98 Description of the rebate/ discount procedure met my needs	N/A	1	2	3	4	5	6	7	1	2	3
Overall											
99 I am satisfied with the information provided by the Web site	N/A	1	2	3	4	5	6	7	1	2	3
100 Overall the Web site provided useful information	N/A	1	2	3	4	5	6	7	1	2	3

Appendix B: Content Preparation Questionnaire for Study 2

PART I: DEMOGRAPHIC INFORMATION

1. Please select your gender.
   - Male
   - Female
2. Please tell us the year you were born. __________
3. Please select your job.
   - Manager
   - Account
   - Sales
   - Office Administrator
   - Engineer
   - Committee
   - Visiting Professor and Scholar
   - Intern student
   - Other
4. Please select your gained (or current) education level.
   - High school
   - Associate College
   - Bachelor
   - Master
   - PhD
   - Other
5. How many years have you been using online shopping? _____
6. If you do not have online shopping experience, have you used the Internet to check the product you are interested in?
   - Yes
   - No

PART II: ONLINE SHOPPING SURVEY

If you have online shopping experience, or you have been looking up products online before, please recall your latest online shopping experience. Please choose *one* product as the assumption background, answer the following questions, and circle the number.

If you do not have online shopping experience, or you have not checked products on the Internet before, please recall your latest shopping experience. Please choose *one* product as the assumption background, answer the following questions, and circle the number.

1. The e-business Web site should describe the size of the product in detail.

Strongly Disagree	Disagree	Slightly Disagree	Neither	Slightly Agree	Agree	Strongly Agree
◊ 1	◊ 2	◊ 3	◊ 4	◊ 5	◊ 6	◊ 7

2. The e-business Web site should provide product accessory information.

Strongly Disagree	Disagree	Slightly Disagree	Neither	Slightly Agree	Agree	Strongly Agree
◊ 1	◊ 2	◊ 3	◊ 4	◊ 5	◊ 6	◊ 7

3. The e-business Web site should provide a detailed description of the product.

Strongly Disagree	Disagree	Slightly Disagree	Neither	Slightly Agree	Agree	Strongly Agree
◊ 1	◊ 2	◊ 3	◊ 4	◊ 5	◊ 6	◊ 7

4. The e-business Web site should provide information about the manufacturer's reputation.

Strongly Disagree	Disagree	Slightly Disagree	Neither	Slightly Agree	Agree	Strongly Agree
◊ 1	◊ 2	◊ 3	◊ 4	◊ 5	◊ 6	◊ 7

5. If the product has multiple style or colors, the e-business Web site should list a picture for every choice.

Strongly Disagree	Disagree	Slightly Disagree	Neither	Slightly Agree	Agree	Strongly Agree
◊ 1	◊ 2	◊ 3	◊ 4	◊ 5	◊ 6	◊ 7

6. The e-business Web site should list the production location of the product.

Strongly Disagree	Disagree	Slightly Disagree	Neither	Slightly Agree	Agree	Strongly Agree
◊ 1	◊ 2	◊ 3	◊ 4	◊ 5	◊ 6	◊ 7

7. It is *not* necessary to list how much money would be saved if customers buy it from the e-business Web site.

Strongly Disagree	Disagree	Slightly Disagree	Neither	Slightly Agree	Agree	Strongly Agree
◊ 1	◊ 2	◊ 3	◊ 4	◊ 5	◊ 6	◊ 7

8. The e-business Web site should describe the smell or flavor of the product, if applicable.

Strongly Disagree	Disagree	Slightly Disagree	Neither	Slightly Agree	Agree	Strongly Agree
◊ 1	◊ 2	◊ 3	◊ 4	◊ 5	◊ 6	◊ 7

9. The e-business Web site does *not* need to describe the appearance of the product.

Strongly Disagree	Disagree	Slightly Disagree	Neither	Slightly Agree	Agree	Strongly Agree
◊ 1	◊ 2	◊ 3	◊ 4	◊ 5	◊ 6	◊ 7

10. The e-business Web site should provide photos of product from different viewpoints directions.

Strongly Disagree	Disagree	Slightly Disagree	Neither	Slightly Agree	Agree	Strongly Agree
◊ 1	◊ 2	◊ 3	◊ 4	◊ 5	◊ 6	◊ 7

11. The e-business Web site should tell me the specialty of the product.

Strongly Disagree	Disagree	Slightly Disagree	Neither	Slightly Agree	Agree	Strongly Agree
◊ 1	◊ 2	◊ 3	◊ 4	◊ 5	◊ 6	◊ 7

12. The e-business Web site should tell me how much money I could save if I purchase the product on their Web site.

Strongly Disagree	Disagree	Slightly Disagree	Neither	Slightly Agree	Agree	Strongly Agree
◊ 1	◊ 2	◊ 3	◊ 4	◊ 5	◊ 6	◊ 7

13. The e-business Web site should list the product's components.

Strongly Disagree	Disagree	Slightly Disagree	Neither	Slightly Agree	Agree	Strongly Agree
◊ 1	◊ 2	◊ 3	◊ 4	◊ 5	◊ 6	◊ 7

14. The e-business Web site should provide the product's technical information.

Strongly Disagree	Disagree	Slightly Disagree	Neither	Slightly Agree	Agree	Strongly Agree
◊ 1	◊ 2	◊ 3	◊ 4	◊ 5	◊ 6	◊ 7

15. The e-business Web site should provide the sell or expiration date of the product, if applicable.

Strongly Disagree	Disagree	Slightly Disagree	Neither	Slightly Agree	Agree	Strongly Agree
◊ 1	◊ 2	◊ 3	◊ 4	◊ 5	◊ 6	◊ 7

16. The e-business Web site should list the manufacturer of the product.

Strongly Disagree	Disagree	Slightly Disagree	Neither	Slightly Agree	Agree	Strongly Agree
◊ 1	◊ 2	◊ 3	◊ 4	◊ 5	◊ 6	◊ 7

17. The e-business Web site should tell me the benefit of the product.

Strongly Disagree	Disagree	Slightly Disagree	Neither	Slightly Agree	Agree	Strongly Agree
◊ 1	◊ 2	◊ 3	◊ 4	◊ 5	◊ 6	◊ 7

18. The e-business Web site should provide price increment/decrement information about the product.

Strongly Disagree	Disagree	Slightly Disagree	Neither	Slightly Agree	Agree	Strongly Agree
◊ 1	◊ 2	◊ 3	◊ 4	◊ 5	◊ 6	◊ 7

19. The e-business Web site should provide the exact weight or volume information of the product.

Strongly Disagree	Disagree	Slightly Disagree	Neither	Slightly Agree	Agree	Strongly Agree
◊ 1	◊ 2	◊ 3	◊ 4	◊ 5	◊ 6	◊ 7

20. The e-business Web site should list the production materials of the product.

Strongly Disagree	Disagree	Slightly Disagree	Neither	Slightly Agree	Agree	Strongly Agree
◊ 1	◊ 2	◊ 3	◊ 4	◊ 5	◊ 6	◊ 7

21. The e-business Web site should provide a quality certificate for customers.

Strongly Disagree	Disagree	Slightly Disagree	Neither	Slightly Agree	Agree	Strongly Agree
◊ 1	◊ 2	◊ 3	◊ 4	◊ 5	◊ 6	◊ 7

22. The e-business Web site should *not* list characteristics of the product.

Strongly Disagree	Disagree	Slightly Disagree	Neither	Slightly Agree	Agree	Strongly Agree
◊ 1	◊ 2	◊ 3	◊ 4	◊ 5	◊ 6	◊ 7

23. The e-business Web site should provide the model type of the product.

Strongly Disagree	Disagree	Slightly Disagree	Neither	Slightly Agree	Agree	Strongly Agree
◊ 1	◊ 2	◊ 3	◊ 4	◊ 5	◊ 6	◊ 7

24. The e-business Web site should provide the ingredients of the product, if applicable.

Strongly Disagree	Disagree	Slightly Disagree	Neither	Slightly Agree	Agree	Strongly Agree
◊ 1	◊ 2	◊ 3	◊ 4	◊ 5	◊ 6	◊ 7

25. The e-business Web site should describe the color of the product.

Strongly Disagree	Disagree	Slightly Disagree	Neither	Slightly Agree	Agree	Strongly Agree
◊ 1	◊ 2	◊ 3	◊ 4	◊ 5	◊ 6	◊ 7

26. The e-business Web site should describe the feeling or touch of the product to me, if applicable.

Strongly Disagree	Disagree	Slightly Disagree	Neither	Slightly Agree	Agree	Strongly Agree
◊ 1	◊ 2	◊ 3	◊ 4	◊ 5	◊ 6	◊ 7

27. The e-business Web site should provide toll-free phone service information.

Strongly Disagree	Disagree	Slightly Disagree	Neither	Slightly Agree	Agree	Strongly Agree
◊ 1	◊ 2	◊ 3	◊ 4	◊ 5	◊ 6	◊ 7

28. The e-business Web site should inform me of Web site news by e-mail or mail.

Strongly Disagree	Disagree	Slightly Disagree	Neither	Slightly Agree	Agree	Strongly Agree
◊ 1	◊ 2	◊ 3	◊ 4	◊ 5	◊ 6	◊ 7

29. The e-business Web site should provide the estimated time frame when customers require service, such as exchange, return, or repair.

Strongly Disagree	Disagree	Slightly Disagree	Neither	Slightly Agree	Agree	Strongly Agree
◊ 1	◊ 2	◊ 3	◊ 4	◊ 5	◊ 6	◊ 7

30. The e-business Web site should *not* provide price and duration for different shipping choices.

Strongly Disagree	Disagree	Slightly Disagree	Neither	Slightly Agree	Agree	Strongly Agree
◊ 1	◊ 2	◊ 3	◊ 4	◊ 5	◊ 6	◊ 7

31. The e-business Web site should provide news regarding new product arrivals.

Strongly Disagree	Disagree	Slightly Disagree	Neither	Slightly Agree	Agree	Strongly Agree
◊ 1	◊ 2	◊ 3	◊ 4	◊ 5	◊ 6	◊ 7

32. The e-business Web site should provide tracking information for the shipment.

Strongly Disagree	Disagree	Slightly Disagree	Neither	Slightly Agree	Agree	Strongly Agree
◊ 1	◊ 2	◊ 3	◊ 4	◊ 5	◊ 6	◊ 7

33. The e-business Web site should list the cost of services, such as exchange, return, or repair.

Strongly Disagree	Disagree	Slightly Disagree	Neither	Slightly Agree	Agree	Strongly Agree
◊ 1	◊ 2	◊ 3	◊ 4	◊ 5	◊ 6	◊ 7

34. The e-business Web site should clarify the purchase responsibility before customers click "place order."

Strongly Disagree	Disagree	Slightly Disagree	Neither	Slightly Agree	Agree	Strongly Agree
◊ 1	◊ 2	◊ 3	◊ 4	◊ 5	◊ 6	◊ 7

35. The e-business Web site should provide multiple shipping choices with different prices.

Strongly Disagree	Disagree	Slightly Disagree	Neither	Slightly Agree	Agree	Strongly Agree
◊ 1	◊ 2	◊ 3	◊ 4	◊ 5	◊ 6	◊ 7

36. The e-business Web site should clearly list any restrictions on discount and sale.

Strongly Disagree	Disagree	Slightly Disagree	Neither	Slightly Agree	Agree	Strongly Agree
◊ 1	◊ 2	◊ 3	◊ 4	◊ 5	◊ 6	◊ 7

37. It is necessary for the e-business Web site to suggest products that may go well with my purchase.

Strongly Disagree	Disagree	Slightly Disagree	Neither	Slightly Agree	Agree	Strongly Agree
◊ 1	◊ 2	◊ 3	◊ 4	◊ 5	◊ 6	◊ 7

38. The e-business Web site should provide a size chart for convenience, if applicable.

Strongly Disagree	Disagree	Slightly Disagree	Neither	Slightly Agree	Agree	Strongly Agree
◊ 1	◊ 2	◊ 3	◊ 4	◊ 5	◊ 6	◊ 7

39. The e-business Web site should provide categories illustrated by pictures.

Strongly Disagree	Disagree	Slightly Disagree	Neither	Slightly Agree	Agree	Strongly Agree
◊ 1	◊ 2	◊ 3	◊ 4	◊ 5	◊ 6	◊ 7

40. The e-business Web site should provide pictures of operation for products that are difficult to operate.

Strongly Disagree	Disagree	Slightly Disagree	Neither	Slightly Agree	Agree	Strongly Agree
◊ 1	◊ 2	◊ 3	◊ 4	◊ 5	◊ 6	◊ 7

41. The e-business Web site should provide technical support if a customer cannot access the webpages.

Strongly Disagree	Disagree	Slightly Disagree	Neither	Slightly Agree	Agree	Strongly Agree
◊ 1	◊ 2	◊ 3	◊ 4	◊ 5	◊ 6	◊ 7

42. Each type of product should have expert comments or recommendations.

Strongly Disagree	Disagree	Slightly Disagree	Neither	Slightly Agree	Agree	Strongly Agree
◊ 1	◊ 2	◊ 3	◊ 4	◊ 5	◊ 6	◊ 7

43. Ratings from other customers are important to me.

Strongly Disagree	Disagree	Slightly Disagree	Neither	Slightly Agree	Agree	Strongly Agree
◊ 1	◊ 2	◊ 3	◊ 4	◊ 5	◊ 6	◊ 7

44. It is *not* necessary for the e-business Web site to suggest products that may go well with my purchase.

Strongly Disagree	Disagree	Slightly Disagree	Neither	Slightly Agree	Agree	Strongly Agree
◊ 1	◊ 2	◊ 3	◊ 4	◊ 5	◊ 6	◊ 7

45. The e-business Web site should provide price comparisons for both local and online stores.

Strongly Disagree	Disagree	Slightly Disagree	Neither	Slightly Agree	Agree	Strongly Agree
◊ 1	◊ 2	◊ 3	◊ 4	◊ 5	◊ 6	◊ 7

46. The e-business Web site should introduce the brands, especially new brands.

Strongly Disagree	Disagree	Slightly Disagree	Neither	Slightly Agree	Agree	Strongly Agree
◊ 1	◊ 2	◊ 3	◊ 4	◊ 5	◊ 6	◊ 7

47. The e-business Web site should provide printable product manuals for customers, if applicable.

Strongly Disagree	Disagree	Slightly Disagree	Neither	Slightly Agree	Agree	Strongly Agree
◊ 1	◊ 2	◊ 3	◊ 4	◊ 5	◊ 6	◊ 7

48. The e-business Web site should provide photos with a size scale notation.

Strongly Disagree	Disagree	Slightly Disagree	Neither	Slightly Agree	Agree	Strongly Agree
◊ 1	◊ 2	◊ 3	◊ 4	◊ 5	◊ 6	◊ 7

49. The e-business Web site should tell me where I can buy the product if it is sold out.

Strongly Disagree	Disagree	Slightly Disagree	Neither	Slightly Agree	Agree	Strongly Agree
◊ 1	◊ 2	◊ 3	◊ 4	◊ 5	◊ 6	◊ 7

50. There should be a search bar on every webpage of the e-business Web site.

Strongly Disagree	Disagree	Slightly Disagree	Neither	Slightly Agree	Agree	Strongly Agree
◊ 1	◊ 2	◊ 3	◊ 4	◊ 5	◊ 6	◊ 7

51. The e-business Web site should provide a site map for searching convenience.

Strongly Disagree	Disagree	Slightly Disagree	Neither	Slightly Agree	Agree	Strongly Agree
◊ 1	◊ 2	◊ 3	◊ 4	◊ 5	◊ 6	◊ 7

52. The e-business Web site should provide a customizing function.

Strongly Disagree	Disagree	Slightly Disagree	Neither	Slightly Agree	Agree	Strongly Agree
◊ 1	◊ 2	◊ 3	◊ 4	◊ 5	◊ 6	◊ 7

53. The contact information of the Web site should be clearly listed on every webpage.

Strongly Disagree	Disagree	Slightly Disagree	Neither	Slightly Agree	Agree	Strongly Agree
◊ 1	◊ 2	◊ 3	◊ 4	◊ 5	◊ 6	◊ 7

54. The e-business Web site should provide:
   a. Search and category functions by brand or manufacturer

Strongly Disagree	Disagree	Slightly Disagree	Neither	Slightly Agree	Agree	Strongly Agree
◊ 1	◊ 2	◊ 3	◊ 4	◊ 5	◊ 6	◊ 7

   b. Search and category functions by price range

Strongly Disagree	Disagree	Slightly Disagree	Neither	Slightly Agree	Agree	Strongly Agree
◊ 1	◊ 2	◊ 3	◊ 4	◊ 5	◊ 6	◊ 7

   c. Search and category functions by price

Strongly Disagree	Disagree	Slightly Disagree	Neither	Slightly Agree	Agree	Strongly Agree
◊ 1	◊ 2	◊ 3	◊ 4	◊ 5	◊ 6	◊ 7

d. Search and category functions by product characteristics (e.g., color, size)

Strongly Disagree	Disagree	Slightly Disagree	Neither	Slightly Agree	Agree	Strongly Agree
◊ 1	◊ 2	◊ 3	◊ 4	◊ 5	◊ 6	◊ 7

e. Search and category functions by discount information

Strongly Disagree	Disagree	Slightly Disagree	Neither	Slightly Agree	Agree	Strongly Agree
◊ 1	◊ 2	◊ 3	◊ 4	◊ 5	◊ 6	◊ 7

f. Search and category functions by customer rating

Strongly Disagree	Disagree	Slightly Disagree	Neither	Slightly Agree	Agree	Strongly Agree
◊ 1	◊ 2	◊ 3	◊ 4	◊ 5	◊ 6	◊ 7

55. The e-business Web site should provide similar products to my described merchandise.

Strongly Disagree	Disagree	Slightly Disagree	Neither	Slightly Agree	Agree	Strongly Agree
◊ 1	◊ 2	◊ 3	◊ 4	◊ 5	◊ 6	◊ 7

56. The e-business Web site should retain the information about the products I just viewed.

Strongly Disagree	Disagree	Slightly Disagree	Neither	Slightly Agree	Agree	Strongly Agree
◊ 1	◊ 2	◊ 3	◊ 4	◊ 5	◊ 6	◊ 7

57. The e-business Web site should keep all my shopping records.

Strongly Disagree	Disagree	Slightly Disagree	Neither	Slightly Agree	Agree	Strongly Agree
◊ 1	◊ 2	◊ 3	◊ 4	◊ 5	◊ 6	◊ 7

58. The e-business Web site should provide links to other Web sites, such as manufacturer Web sites.

Strongly Disagree	Disagree	Slightly Disagree	Neither	Slightly Agree	Agree	Strongly Agree
◊ 1	◊ 2	◊ 3	◊ 4	◊ 5	◊ 6	◊ 7

59. The e-business Web site should provide example for video operation products that are difficult to operate.

Strongly Disagree	Disagree	Slightly Disagree	Neither	Slightly Agree	Agree	Strongly Agree
◊ 1	◊ 2	◊ 3	◊ 4	◊ 5	◊ 6	◊ 7

60. Photos provided by other customers would help me to get to know the product.

Strongly Disagree	Disagree	Slightly Disagree	Neither	Slightly Agree	Agree	Strongly Agree
◊ 1	◊ 2	◊ 3	◊ 4	◊ 5	◊ 6	◊ 7

61. The e-business Web site should provide a forum for customers to discuss products.

Strongly Disagree	Disagree	Slightly Disagree	Neither	Slightly Agree	Agree	Strongly Agree
◊ 1	◊ 2	◊ 3	◊ 4	◊ 5	◊ 6	◊ 7

62. Comments from other customers are important to me.

Strongly Disagree	Disagree	Slightly Disagree	Neither	Slightly Agree	Agree	Strongly Agree
◊ 1	◊ 2	◊ 3	◊ 4	◊ 5	◊ 6	◊ 7

63. It is important to know membership responsibilities, annual fees, and benefits before subscribing.

Strongly Disagree	Disagree	Slightly Disagree	Neither	Slightly Agree	Agree	Strongly Agree
◊ 1	◊ 2	◊ 3	◊ 4	◊ 5	◊ 6	◊ 7

64. The e-business Web site should provide information about protecting customer privacy, and safety.

Strongly Disagree	Disagree	Slightly Disagree	Neither	Slightly Agree	Agree	Strongly Agree
◊ 1	◊ 2	◊ 3	◊ 4	◊ 5	◊ 6	◊ 7

65. The e-business Web site should provide information that promise the safety of credit cards.

Strongly Disagree	Disagree	Slightly Disagree	Neither	Slightly Agree	Agree	Strongly Agree
◊ 1	◊ 2	◊ 3	◊ 4	◊ 5	◊ 6	◊ 7

66. What kind of product did you assume as the subject just now during the survey?
    - Books and magazines
    - Clothing and clothing accessories
    - Computer hardware
    - Computer software
    - Drugs, health aids, and beauty aids
    - Electronics and appliance
    - Food, beer, and wine
    - Furniture and home furnishing
    - Music and videos
    - Office equipment and supplies
    - Sporting goods
    - Toys, hobby goods, and games
    - Other ____________________

Appendix C: Web-Based Survey of Content Information on E-Commerce Web Sites for Study 3

1. Name: ___________
2. Gender: ◊ Female ◊ Male
3. Age: ___________
4. Education status: ◊ Undergraduate ◊ Graduate
5. Major:

- ◊ Engineering
- ◊ Science
- ◊ Other (Please indicate) _____________
- ◊ Humanities

Questions 6 through 33 contain statements regarding information content that appears on the majority of e-commerce Web sites. Assume you are going to buy a portable electronic product (e.g., MP3 players, digital cameras, and laptop computers) online, please mark the response that best reflects *your* opinion based on your experience with online shopping. Please read each question carefully, but do not spend too much time in choosing your response. Thank you for your cooperation!

6. When choosing a portable electronic product, I pay a lot of attention to the color of the product.

Strongly Disagree	Disagree	Slightly Disagree	Neither	Slightly Agree	Agree	Strongly Agree
◊ 1	◊ 2	◊ 3	◊ 4	◊ 5	◊ 6	◊ 7

7. When choosing a portable electronic product, I pay a lot of attention to the weight of the product.

Strongly Disagree	Disagree	Slightly Disagree	Neither	Slightly Agree	Agree	Strongly Agree
◊ 1	◊ 2	◊ 3	◊ 4	◊ 5	◊ 6	◊ 7

8. When choosing a portable electronic product, I pay a lot of attention to the size of the product.

Strongly Disagree	Disagree	Slightly Disagree	Neither	Slightly Agree	Agree	Strongly Agree
◊ 1	◊ 2	◊ 3	◊ 4	◊ 5	◊ 6	◊ 7

9. When choosing a portable electronic product, I pay a lot of attention to the material that the product is made of.

Strongly Disagree	Disagree	Slightly Disagree	Neither	Slightly Agree	Agree	Strongly Agree
◊ 1	◊ 2	◊ 3	◊ 4	◊ 5	◊ 6	◊ 7

10. When choosing a portable electronic product, I pay a lot of attention to in which country the product was made.

Strongly Disagree	Disagree	Slightly Disagree	Neither	Slightly Agree	Agree	Strongly Agree
◊ 1	◊ 2	◊ 3	◊ 4	◊ 5	◊ 6	◊ 7

11. When choosing a portable electronic product, I pay a lot of attention to the product composition and ancillary items included with the product.

Strongly Disagree	Disagree	Slightly Disagree	Neither	Slightly Agree	Agree	Strongly Agree
◊ 1	◊ 2	◊ 3	◊ 4	◊ 5	◊ 6	◊ 7

12. When choosing a portable electronic product, I pay a lot of attention to the new concepts and features introduced in and/or advantages offered by the product, and new and contemporary technology utilized in the product.

Strongly Disagree	Disagree	Slightly Disagree	Neither	Slightly Agree	Agree	Strongly Agree
◊ 1	◊ 2	◊ 3	◊ 4	◊ 5	◊ 6	◊ 7

13. When choosing a portable electronic product, I pay a lot of attention to the advanced and sophisticated technical skills used to engineer and manufacture the product.

Strongly Disagree	Disagree	Slightly Disagree	Neither	Slightly Agree	Agree	Strongly Agree
◊ 1	◊ 2	◊ 3	◊ 4	◊ 5	◊ 6	◊ 7

14. When deciding which portable electronic product to buy, I place a lot of emphasis on the product's price.

Strongly Disagree	Disagree	Slightly Disagree	Neither	Slightly Agree	Agree	Strongly Agree
◊ 1	◊ 2	◊ 3	◊ 4	◊ 5	◊ 6	◊ 7

15. When deciding which portable electronic product to buy, I place a lot of emphasis on how well the product can retain its value.

Strongly Disagree	Disagree	Slightly Disagree	Neither	Slightly Agree	Agree	Strongly Agree
◊ 1	◊ 2	◊ 3	◊ 4	◊ 5	◊ 6	◊ 7

16. When deciding which portable electronic product to buy, I place a lot of emphasis on the product's capability of meeting my needs relative to its price.

Strongly Disagree	Disagree	Slightly Disagree	Neither	Slightly Agree	Agree	Strongly Agree
◊ 1	◊ 2	◊ 3	◊ 4	◊ 5	◊ 6	◊ 7

17. When deciding which portable electronic product to buy, I do *not* care how much the product weighs.*

Strongly Disagree	Disagree	Slightly Disagree	Neither	Slightly Agree	Agree	Strongly Agree
◊ 1	◊ 2	◊ 3	◊ 4	◊ 5	◊ 6	◊ 7

18. When deciding which portable electronic product to buy, I place a lot of emphasis on what the product does, and how well the product does what it is designed to do in comparison to competing products.

Strongly Disagree	Disagree	Slightly Disagree	Neither	Slightly Agree	Agree	Strongly Agree
◊ 1	◊ 2	◊ 3	◊ 4	◊ 5	◊ 6	◊ 7

19. When deciding which portable electronic product to buy, I place a lot of emphasis on how easy the product is to use.

Strongly Disagree	Disagree	Slightly Disagree	Neither	Slightly Agree	Agree	Strongly Agree
◊ 1	◊ 2	◊ 3	◊ 4	◊ 5	◊ 6	◊ 7

20. When deciding which portable electronic product to buy, I place a lot of emphasis on whether the product is cost-effective to buy.

Strongly Disagree	Disagree	Slightly Disagree	Neither	Slightly Agree	Agree	Strongly Agree
◊ 1	◊ 2	◊ 3	◊ 4	◊ 5	◊ 6	◊ 7

21. When deciding which portable electronic product to buy, I place a lot of emphasis on the product's characteristics that distinguish it from competing products based on an objective evaluation of workmanship, engineering, durability, and excellence of materials.

Strongly Disagree	Disagree	Slightly Disagree	Neither	Slightly Agree	Agree	Strongly Agree
◊ 1	◊ 2	◊ 3	◊ 4	◊ 5	◊ 6	◊ 7

22. When deciding which portable electronic product to buy, I place a lot of emphasis on the warranties that accompany the product during a specific period of time.

Strongly Disagree	Disagree	Slightly Disagree	Neither	Slightly Agree	Agree	Strongly Agree
◊ 1	◊ 2	◊ 3	◊ 4	◊ 5	◊ 6	◊ 7

23. When deciding which portable electronic product to buy, I place a lot of emphasis on the post-purchase technical and/or service assistance during a specific period of time for a given product.

Strongly Disagree	Disagree	Slightly Disagree	Neither	Slightly Agree	Agree	Strongly Agree
◊ 1	◊ 2	◊ 3	◊ 4	◊ 5	◊ 6	◊ 7

24. When deciding which portable electronic product to buy, I do *not* take the size of the product into consideration.*

Strongly Disagree	Disagree	Slightly Disagree	Neither	Slightly Agree	Agree	Strongly Agree
◊ 1	◊ 2	◊ 3	◊ 4	◊ 5	◊ 6	◊ 7

25. The information below has a high weighting in my decision of which portable electronic product to buy.

    Data presented to compare the product with other competing products, such as industry credentials, sales rankings, technical inspections, etc.

Strongly Disagree	Disagree	Slightly Disagree	Neither	Slightly Agree	Agree	Strongly Agree
◊ 1	◊ 2	◊ 3	◊ 4	◊ 5	◊ 6	◊ 7

26. When buying a portable electronic product online, I pay a lot of attention to the conditions under which returns and exchanges can be made.

Strongly Disagree	Disagree	Slightly Disagree	Neither	Slightly Agree	Agree	Strongly Agree
◊ 1	◊ 2	◊ 3	◊ 4	◊ 5	◊ 6	◊ 7

27. When buying a portable electronic product online, I pay a lot of attention to how to return the product and/or make an exchange if I am not satisfied with the product.

Strongly Disagree	Disagree	Slightly Disagree	Neither	Slightly Agree	Agree	Strongly Agree
◊ 1	◊ 2	◊ 3	◊ 4	◊ 5	◊ 6	◊ 7

28. When buying a portable electronic product online, I pay a lot of attention to whether I will get refunded after I return a product.

Strongly Disagree	Disagree	Slightly Disagree	Neither	Slightly Agree	Agree	Strongly Agree
◊ 1	◊ 2	◊ 3	◊ 4	◊ 5	◊ 6	◊ 7

29. When buying a portable electronic product online, I do *not* check in which country it was manufactured.*

Strongly Disagree	Disagree	Slightly Disagree	Neither	Slightly Agree	Agree	Strongly Agree
◊ 1	◊ 2	◊ 3	◊ 4	◊ 5	◊ 6	◊ 7

30. When buying a portable electronic product online, I place a lot of emphasis on the security of the online transaction.

Strongly Disagree	Disagree	Slightly Disagree	Neither	Slightly Agree	Agree	Strongly Agree
◊ 1	◊ 2	◊ 3	◊ 4	◊ 5	◊ 6	◊ 7

31. When buying a portable electronic product online, I place a lot of emphasis on the privacy policies of the e-commerce Web site.

Strongly Disagree	Disagree	Slightly Disagree	Neither	Slightly Agree	Agree	Strongly Agree
◊ 1	◊ 2	◊ 3	◊ 4	◊ 5	◊ 6	◊ 7

32. When buying a portable electronic product online, I pay a lot of attention to information about how to contact, if the need arises, the representatives of the online retailer.

Strongly Disagree	Disagree	Slightly Disagree	Neither	Slightly Agree	Agree	Strongly Agree
◊ 1	◊ 2	◊ 3	◊ 4	◊ 5	◊ 6	◊ 7

33. When deciding which portable electronic product to buy, I pay a lot of attention to the product's safety features and the possible hazard/harm when using the product.

Strongly Disagree	Disagree	Slightly Disagree	Neither	Slightly Agree	Agree	Strongly Agree
◊ 1	◊ 2	◊ 3	◊ 4	◊ 5	◊ 6	◊ 7

* Items reversely scored.

Appendix D: Content Preparation Questionnaire for Information Appliances for Study 4

PART I: DEMOGRAPHIC QUESTIONS

This is an anonymous questionnaire. Please fill out *every* question in the following questionnaire. Thanks!

1. What is your gender?
   - Female
   - Male
2. Please tell us in which year you were born_______________.
3. Please select your job category:
   - Manager
   - Sales and Marketing
   - Office Administrator
   - Engineer
   - Technician and workers
   - Visiting professor and scholar
   - Student
   - Other
4. Please select your gained (or current) education level:
   - High school
   - Associate College
   - Bachelor
   - Master
   - PhD
   - Other
5. How many cell phones have you used before?
   - 0
   - 1
   - 2
   - More than 2

6. In which year did you start using a cell phone? ____________
7. What are the most important factors influencing your purchase of a cell phone? Please mark from 1 (most important) to 6 (most unimportant).
   _____ Brand
   _____ Quality
   _____ Good usability design
   _____ Price
   _____ Fashion design
   _____ Multiple functions

PART II: CONTENT QUESTIONS

Please fill out *every* question in the following questionnaire according to your experience of using a cell phone. Please circle your answer (limited to one for each question).

1. The cell phone should have an MP3 player function.

Strongly Disagree	Disagree	Slightly Disagree	Neither	Slightly Agree	Agree	Strongly Agree
◊ 1	◊ 2	◊ 3	◊ 4	◊ 5	◊ 6	◊ 7

2. The cell phone should have a digital camera function.

Strongly Disagree	Disagree	Slightly Disagree	Neither	Slightly Agree	Agree	Strongly Agree
◊ 1	◊ 2	◊ 3	◊ 4	◊ 5	◊ 6	◊ 7

3. The cell phone should have a sequential shooting function.

Strongly Disagree	Disagree	Slightly Disagree	Neither	Slightly Agree	Agree	Strongly Agree
◊ 1	◊ 2	◊ 3	◊ 4	◊ 5	◊ 6	◊ 7

4. The cell phone should have a digital video function.

Strongly Disagree	Disagree	Slightly Disagree	Neither	Slightly Agree	Agree	Strongly Agree
◊ 1	◊ 2	◊ 3	◊ 4	◊ 5	◊ 6	◊ 7

5. The cell phone should have a mobile television function.

Strongly Disagree	Disagree	Slightly Disagree	Neither	Slightly Agree	Agree	Strongly Agree
◊ 1	◊ 2	◊ 3	◊ 4	◊ 5	◊ 6	◊ 7

6. The cell phone should have a mobile game function.

Strongly Disagree	Disagree	Slightly Disagree	Neither	Slightly Agree	Agree	Strongly Agree
◊ 1	◊ 2	◊ 3	◊ 4	◊ 5	◊ 6	◊ 7

7. The cell phone should have a calendar function.

Strongly Disagree	Disagree	Slightly Disagree	Neither	Slightly Agree	Agree	Strongly Agree
◊ 1	◊ 2	◊ 3	◊ 4	◊ 5	◊ 6	◊ 7

8. The cell phone should have an e-book function.

Strongly Disagree	Disagree	Slightly Disagree	Neither	Slightly Agree	Agree	Strongly Agree
◊ 1	◊ 2	◊ 3	◊ 4	◊ 5	◊ 6	◊ 7

9. The cell phone should have an e-dictionary function.

Strongly Disagree	Disagree	Slightly Disagree	Neither	Slightly Agree	Agree	Strongly Agree
◊ 1	◊ 2	◊ 3	◊ 4	◊ 5	◊ 6	◊ 7

10. The cell phone should have a GPS function.

Strongly Disagree	Disagree	Slightly Disagree	Neither	Slightly Agree	Agree	Strongly Agree
◊ 1	◊ 2	◊ 3	◊ 4	◊ 5	◊ 6	◊ 7

11. The cell phone should have a traffic information function.

Strongly Disagree	Disagree	Slightly Disagree	Neither	Slightly Agree	Agree	Strongly Agree
◊ 1	◊ 2	◊ 3	◊ 4	◊ 5	◊ 6	◊ 7

12. The cell phone should have a customization function.

Strongly Disagree	Disagree	Slightly Disagree	Neither	Slightly Agree	Agree	Strongly Agree
◊ 1	◊ 2	◊ 3	◊ 4	◊ 5	◊ 6	◊ 7

13. It is not necessary for the cell phone to have an MP3 player function.

Strongly Disagree	Disagree	Slightly Disagree	Neither	Slightly Agree	Agree	Strongly Agree
◊ 1	◊ 2	◊ 3	◊ 4	◊ 5	◊ 6	◊ 7

14. The cell phone should have an online surfing function.

Strongly Disagree	Disagree	Slightly Disagree	Neither	Slightly Agree	Agree	Strongly Agree
◊ 1	◊ 2	◊ 3	◊ 4	◊ 5	◊ 6	◊ 7

15. The cell phone should have an instant messenger function.

Strongly Disagree	Disagree	Slightly Disagree	Neither	Slightly Agree	Agree	Strongly Agree
◊ 1	◊ 2	◊ 3	◊ 4	◊ 5	◊ 6	◊ 7

16. The cell phone should have an emergency key.

Strongly Disagree	Disagree	Slightly Disagree	Neither	Slightly Agree	Agree	Strongly Agree
◊ 1	◊ 2	◊ 3	◊ 4	◊ 5	◊ 6	◊ 7

17. The cell phone should have a video phone function.

Strongly Disagree	Disagree	Slightly Disagree	Neither	Slightly Agree	Agree	Strongly Agree
◊ 1	◊ 2	◊ 3	◊ 4	◊ 5	◊ 6	◊ 7

18. The cell phone should have a handwriting input function.

Strongly Disagree	Disagree	Slightly Disagree	Neither	Slightly Agree	Agree	Strongly Agree
◊ 1	◊ 2	◊ 3	◊ 4	◊ 5	◊ 6	◊ 7

19. It is necessary to provide the name of function when you are looking at the main menu.

Strongly Disagree	Disagree	Slightly Disagree	Neither	Slightly Agree	Agree	Strongly Agree
◊ 1	◊ 2	◊ 3	◊ 4	◊ 5	◊ 6	◊ 7

20. It is necessary to provide the name of function when you are looking at the menus of each level.

Strongly Disagree	Disagree	Slightly Disagree	Neither	Slightly Agree	Agree	Strongly Agree
◊ 1	◊ 2	◊ 3	◊ 4	◊ 5	◊ 6	◊ 7

21. It is necessary to provide the icon of function when you are looking at the main menu.

Strongly Disagree	Disagree	Slightly Disagree	Neither	Slightly Agree	Agree	Strongly Agree
◊ 1	◊ 2	◊ 3	◊ 4	◊ 5	◊ 6	◊ 7

22. It is necessary to provide the sequence number of function when you are looking at the main menu.

Strongly Disagree	Disagree	Slightly Disagree	Neither	Slightly Agree	Agree	Strongly Agree
◊ 1	◊ 2	◊ 3	◊ 4	◊ 5	◊ 6	◊ 7

23. The cell phone should provide all options for each function on any menu.

Strongly Disagree	Disagree	Slightly Disagree	Neither	Slightly Agree	Agree	Strongly Agree
◊ 1	◊ 2	◊ 3	◊ 4	◊ 5	◊ 6	◊ 7

24. The cell phone should use a cursor to indicate position on the menu interface.

Strongly Disagree	Disagree	Slightly Disagree	Neither	Slightly Agree	Agree	Strongly Agree
◊ 1	◊ 2	◊ 3	◊ 4	◊ 5	◊ 6	◊ 7

25. The cell phone should provide scroll bar to remind me more information is available.

Strongly Disagree	Disagree	Slightly Disagree	Neither	Slightly Agree	Agree	Strongly Agree
◊ 1	◊ 2	◊ 3	◊ 4	◊ 5	◊ 6	◊ 7

26. The cell phone should provide a sequence number for each function.

Strongly Disagree	Disagree	Slightly Disagree	Neither	Slightly Agree	Agree	Strongly Agree
◊ 1	◊ 2	◊ 3	◊ 4	◊ 5	◊ 6	◊ 7

27. It is necessary to provide current status of the function in use.

Strongly Disagree	Disagree	Slightly Disagree	Neither	Slightly Agree	Agree	Strongly Agree
◊ 1	◊ 2	◊ 3	◊ 4	◊ 5	◊ 6	◊ 7

28. The cell phone should provide information about the current settings of the function in use.

Strongly Disagree	Disagree	Slightly Disagree	Neither	Slightly Agree	Agree	Strongly Agree
◊ 1	◊ 2	◊ 3	◊ 4	◊ 5	◊ 6	◊ 7

29. The cell phone should provide a "back to previous menu" key for the menu of each level.

Strongly Disagree	Disagree	Slightly Disagree	Neither	Slightly Agree	Agree	Strongly Agree
◊ 1	◊ 2	◊ 3	◊ 4	◊ 5	◊ 6	◊ 7

30. It is necessary to provide an indication of current activated key on the screen.

Strongly Disagree	Disagree	Slightly Disagree	Neither	Slightly Agree	Agree	Strongly Agree
◊ 1	◊ 2	◊ 3	◊ 4	◊ 5	◊ 6	◊ 7

31. It is necessary to provide an indication of "back to previous menu" key on the screen.

Strongly Disagree	Disagree	Slightly Disagree	Neither	Slightly Agree	Agree	Strongly Agree
◊ 1	◊ 2	◊ 3	◊ 4	◊ 5	◊ 6	◊ 7

32. The cell phone should provide a confirmation indicator.

Strongly Disagree	Disagree	Slightly Disagree	Neither	Slightly Agree	Agree	Strongly Agree
◊ 1	◊ 2	◊ 3	◊ 4	◊ 5	◊ 6	◊ 7

33. It is necessary to indicate which keys are in use on the cell phone interface.

Strongly Disagree	Disagree	Slightly Disagree	Neither	Slightly Agree	Agree	Strongly Agree
◊ 1	◊ 2	◊ 3	◊ 4	◊ 5	◊ 6	◊ 7

34. The cell phone should have a dual-time zone function.

Strongly Disagree	Disagree	Slightly Disagree	Neither	Slightly Agree	Agree	Strongly Agree
◊ 1	◊ 2	◊ 3	◊ 4	◊ 5	◊ 6	◊ 7

35. The cell phone should be able to provide an alert message for breakout incidents.

Strongly Disagree	Disagree	Slightly Disagree	Neither	Slightly Agree	Agree	Strongly Agree
◊ 1	◊ 2	◊ 3	◊ 4	◊ 5	◊ 6	◊ 7

36. It is necessary to provide notification of message status.

Strongly Disagree	Disagree	Slightly Disagree	Neither	Slightly Agree	Agree	Strongly Agree
◊ 1	◊ 2	◊ 3	◊ 4	◊ 5	◊ 6	◊ 7

37. It is necessary to provide an icon for message box status.

Strongly Disagree	Disagree	Slightly Disagree	Neither	Slightly Agree	Agree	Strongly Agree
◊ 1	◊ 2	◊ 3	◊ 4	◊ 5	◊ 6	◊ 7

38. It is necessary to provide an icon for voice mail status.

Strongly Disagree	Disagree	Slightly Disagree	Neither	Slightly Agree	Agree	Strongly Agree
◊ 1	◊ 2	◊ 3	◊ 4	◊ 5	◊ 6	◊ 7

39. It is necessary to provide an icon for “memo” status.

Strongly Disagree	Disagree	Slightly Disagree	Neither	Slightly Agree	Agree	Strongly Agree
◊ 1	◊ 2	◊ 3	◊ 4	◊ 5	◊ 6	◊ 7

40. It is necessary to provide an icon for reminder messages.

Strongly Disagree	Disagree	Slightly Disagree	Neither	Slightly Agree	Agree	Strongly Agree
◊ 1	◊ 2	◊ 3	◊ 4	◊ 5	◊ 6	◊ 7

41. It is necessary to provide an animation for power on/off.

Strongly Disagree	Disagree	Slightly Disagree	Neither	Slightly Agree	Agree	Strongly Agree
◊ 1	◊ 2	◊ 3	◊ 4	◊ 5	◊ 6	◊ 7

42. It is necessary to provide the name of files stored in the cell phone.

Strongly Disagree	Disagree	Slightly Disagree	Neither	Slightly Agree	Agree	Strongly Agree
◊ 1	◊ 2	◊ 3	◊ 4	◊ 5	◊ 6	◊ 7

43. It is necessary to show the properties of files stored in the cell phone.

Strongly Disagree	Disagree	Slightly Disagree	Neither	Slightly Agree	Agree	Strongly Agree
◊ 1	◊ 2	◊ 3	◊ 4	◊ 5	◊ 6	◊ 7

44. It is necessary to show the size of files stored in the cell phone.

Strongly Disagree	Disagree	Slightly Disagree	Neither	Slightly Agree	Agree	Strongly Agree
◊ 1	◊ 2	◊ 3	◊ 4	◊ 5	◊ 6	◊ 7

45. It is necessary to provide the size of photos taken or stored in the cell phone.

Strongly Disagree	Disagree	Slightly Disagree	Neither	Slightly Agree	Agree	Strongly Agree
◊ 1	◊ 2	◊ 3	◊ 4	◊ 5	◊ 6	◊ 7

46. The cell phone should provide information about available storage and free space on the cell phone.

Strongly Disagree	Disagree	Slightly Disagree	Neither	Slightly Agree	Agree	Strongly Agree
◊ 1	◊ 2	◊ 3	◊ 4	◊ 5	◊ 6	◊ 7

47. The cell phone should provide a recycle bin to store recently deleted files.

Strongly Disagree	Disagree	Slightly Disagree	Neither	Slightly Agree	Agree	Strongly Agree
◊ 1	◊ 2	◊ 3	◊ 4	◊ 5	◊ 6	◊ 7

48. It is necessary to provide a search function for files stored in the cell phone.

Strongly Disagree	Disagree	Slightly Disagree	Neither	Slightly Agree	Agree	Strongly Agree
◊ 1	◊ 2	◊ 3	◊ 4	◊ 5	◊ 6	◊ 7

49. The cell phone should provide a "search by name" function for finding contacts.

Strongly Disagree	Disagree	Slightly Disagree	Neither	Slightly Agree	Agree	Strongly Agree
◊ 1	◊ 2	◊ 3	◊ 4	◊ 5	◊ 6	◊ 7

50. The cell phone should provide a "search by number" function for finding contacts.

Strongly Disagree	Disagree	Slightly Disagree	Neither	Slightly Agree	Agree	Strongly Agree
◊ 1	◊ 2	◊ 3	◊ 4	◊ 5	◊ 6	◊ 7

51. The cell phone should provide a "search by initial" function.

Strongly Disagree	Disagree	Slightly Disagree	Neither	Slightly Agree	Agree	Strongly Agree
◊ 1	◊ 2	◊ 3	◊ 4	◊ 5	◊ 6	◊ 7

52. It is necessary to show the current input method especially when users are inputting Chinese.

Strongly Disagree	Disagree	Slightly Disagree	Neither	Slightly Agree	Agree	Strongly Agree
◊ 1	◊ 2	◊ 3	◊ 4	◊ 5	◊ 6	◊ 7

53. It is necessary to show what content has been input.

Strongly Disagree	Disagree	Slightly Disagree	Neither	Slightly Agree	Agree	Strongly Agree
◊ 1	◊ 2	◊ 3	◊ 4	◊ 5	◊ 6	◊ 7

54. It is necessary to provide for the input of "pinyin" letters.

Strongly Disagree	Disagree	Slightly Disagree	Neither	Slightly Agree	Agree	Strongly Agree
◊ 1	◊ 2	◊ 3	◊ 4	◊ 5	◊ 6	◊ 7

55. It is necessary to provide the manufacturer's information for cell phone users.

Strongly Disagree	Disagree	Slightly Disagree	Neither	Slightly Agree	Agree	Strongly Agree
◊ 1	◊ 2	◊ 3	◊ 4	◊ 5	◊ 6	◊ 7

56. It is necessary to provide the contact information of the signal carrier.

Strongly Disagree	Disagree	Slightly Disagree	Neither	Slightly Agree	Agree	Strongly Agree
◊ 1	◊ 2	◊ 3	◊ 4	◊ 5	◊ 6	◊ 7

57. The cell phone should provide help information for each function.

Strongly Disagree	Disagree	Slightly Disagree	Neither	Slightly Agree	Agree	Strongly Agree
◊ 1	◊ 2	◊ 3	◊ 4	◊ 5	◊ 6	◊ 7

58. The cell phone should display the current input method.

Strongly Disagree	Disagree	Slightly Disagree	Neither	Slightly Agree	Agree	Strongly Agree
◊ 1	◊ 2	◊ 3	◊ 4	◊ 5	◊ 6	◊ 7

59. The cell phone should provide service information for manufacturer and signal carrier.

Strongly Disagree	Disagree	Slightly Disagree	Neither	Slightly Agree	Agree	Strongly Agree
◊ 1	◊ 2	◊ 3	◊ 4	◊ 5	◊ 6	◊ 7

60. The cell phone should provide the time of a missed call.

Strongly Disagree	Disagree	Slightly Disagree	Neither	Slightly Agree	Agree	Strongly Agree
◊ 1	◊ 2	◊ 3	◊ 4	◊ 5	◊ 6	◊ 7

61. The cell phone should provide the number of missed calls.

Strongly Disagree	Disagree	Slightly Disagree	Neither	Slightly Agree	Agree	Strongly Agree
◊ 1	◊ 2	◊ 3	◊ 4	◊ 5	◊ 6	◊ 7

62. It is necessary to provide length of each call to the cell phone user.

Strongly Disagree	Disagree	Slightly Disagree	Neither	Slightly Agree	Agree	Strongly Agree
◊ 1	◊ 2	◊ 3	◊ 4	◊ 5	◊ 6	◊ 7

63. It is necessary to provide number of connections for each number.

Strongly Disagree	Disagree	Slightly Disagree	Neither	Slightly Agree	Agree	Strongly Agree
◊ 1	◊ 2	◊ 3	◊ 4	◊ 5	◊ 6	◊ 7

64. It is necessary to provide a "sent" box for recent sent messages.

Strongly Disagree	Disagree	Slightly Disagree	Neither	Slightly Agree	Agree	Strongly Agree
◊ 1	◊ 2	◊ 3	◊ 4	◊ 5	◊ 6	◊ 7

65. According to the above questions, do you think the amount of cell phone content will influence your satisfaction with the cell phone?

Strongly Disagree	Disagree	Slightly Disagree	Neither	Slightly Agree	Agree	Strongly Agree
◊ 1	◊ 2	◊ 3	◊ 4	◊ 5	◊ 6	◊ 7

66. According to the above questions, do you think the amount of cell phone content will influence your speed operating the cell phone?

Strongly Disagree	Disagree	Slightly Disagree	Neither	Slightly Agree	Agree	Strongly Agree
◊ 1	◊ 2	◊ 3	◊ 4	◊ 5	◊ 6	◊ 7

67. According to the above questions, do you think the amount of cell phone content will influence your error rate?

Strongly Disagree	Disagree	Slightly Disagree	Neither	Slightly Agree	Agree	Strongly Agree
◊ 1	◊ 2	◊ 3	◊ 4	◊ 5	◊ 6	◊ 7

68. Do you think that the amount of current cell phone content is enough?

Strongly Disagree	Disagree	Slightly Disagree	Neither	Slightly Agree	Agree	Strongly Agree
◊ 1	◊ 2	◊ 3	◊ 4	◊ 5	◊ 6	◊ 7

References

Abbott, M., Chiang, K. P., Hwang, Y. S., Paquin, J., and Zwick, D. 2000. The process of online store loyalty formation. *Advance in Consumer Research* 27:145–50.

Abram, S. 2003. Why should I care about standards? (Information Trends). *Information Outlook* 7:21.

Ackerman, M. S., Cranor, L. F., and Reagle, J. 1999. Privacy in e-commerce: Examining user scenarios and privacy preferences. Paper presented at 1st ACM Conference on Electronic Commerce, Denver, CO.

ACNielsen. 2005. Global consumer attitudes towards online shopping. http://www2.acnielsen.com/reports/documents/2005_cc_onlineshopping.pdf (Accessed July 1, 2006).

Aladwani, A. M., and Palvia, P. 2002. Developing and validating an instrument for measuring user-perceived web quality. *Information and Management* 39:467–76.

Alexander, J. E., and Tate, M. A. 1999. *Web wisdom: How to evaluate and create information quality on the Web.* Mahwah, NJ: Lawrence Erlbaum.

Anderson, S. P., and Renault, R. 2006. Advertising content. *American Economic Review* 96(1):93–113.

Baierova, P., Tate, M., and Hope, B. 2003. The impact of purpose for web use on user preferences for web design features. Paper presented at 7th Pacific Asia Conference on Information Systems, Adelaide, South Australia.

Barnes, S., and Vidgen, R. 2001. Assessing the quality of auction Web sites. Paper presented at 34th Hawaii International Conference on System Sciences, Maui, HI.

Bayles, D. 2001. *E-commerce logistics and fulfillment: Delivering the goods.* Upper Saddle River, NJ: Prentice Hall PTR.

Beijing Review. 1978. Communique of the third plenum of the eleventh party congress. December 29.

Berr, J. 2006. Amazon buys shopbop. http://www.thestreet.com/_cnet/tech/internet/10270510.html?cm_ven=CNET&cm_cat=FREE&cm_ite=NA (Accessed May 18, 2006).

Bevan, N. 1999. Quality in use: Meeting user needs for quality. *The Journal of Systems and Software* 49 (1):89–96.

Bin, Q., Chen, S.-J., and Sun, S. Q. 2003. Cultural differences in e-commerce: A comparison between the US and China. *Journal of Global Information Management* 11:48–55.

Bloch, P. H., and Richins, M. L. 1983. A theoretical model for the study of product importance perceptions. *Journal of Marketing* 47(Summer):69–81.

Business Wire. 1999. Study finds lack of color consistency hampers electronic commerce: Cyber Dialogue reports consumer awareness of monitor color variance. http://findarticles.com/p/articles/mi_m0EIN/is_1999_April_6/ai_54296014 (Accessed December 15, 2006).

Buys, M., and Brown, I. 2004. Customer satisfaction with internet banking Web sites: An empirical test and validation of a measuring instrument. Presented at the 2004 Annual Research Conference of the South African Institute of Computer Scientists and Information Technologists on IT Research in Developing Countries, Stellenbosch, Western Cape, South Africa.

Carlson, E. R. 1962. Generality of order of concept attainment. *Psychological Reports* 10:375–80.

Chan, E. S. K., and Swatman, P. M. C. 2002. Web content and design: A review of e-commerce/e-business program sites. Paper presented at 13th Australasian Conference of Information Systems, Melbourne, Australia.

Chapanis, A. 1981. Humanizing computers. Presented at the 1981 National Telecommunications Conference, NY.

Cheng, H. 1994. Reflections of cultural values: A content analysis of Chinese magazine advertisements from 1982 and 1992. *International Journal of Advertising* 13:167–83.

Cheng, H., and Schweitzer, J. C. 1996. Cultural values reflected in Chinese and US television commercials. *Journal of Advertising Research* 36:27–45.

Cheung, C. M., and Lee, M. K. 2005. Consumer satisfaction with internet shopping: A research framework and propositions for future research. Paper presented at 7th International Conference on Electronic Commerce (ICEC 2005), Xi'an, China.

China Internet Network Information Center (CNNIC). 2006. 17th statistical survey report on the Internet development in China. http://www.cnnic.net.cn/en/index/0O/02/index.htm (Accessed July 1, 2006).

Chiu, L. H. 1972. A cross-cultural comparison of cognitive styles in Chinese and American children. *International Journal of Psychology* 7:235–42.

Cho, N., and Park, S. 2001. Development of electronic commerce user-consumer satisfaction index (ECUSI) for Internet shopping. *Industrial Management & Data Systems* 101(8):400–406.

Choi, Y., and Miracle, G. E. 2004. The effectiveness of comparative advertising in Korea and the United States: A cross-cultural and individual-level analysis. *Journal of Advertising* 33(4):75–87.

Choong, Y. Y. 1996. Design of computer interfaces for the Chinese population. PhD diss., Purdue University.

Choong, Y. Y., Plocher, T., and Rau, P. P. 2005. Cross-cultural Web design. In *Handbook of human factors in Web design,* ed. R. W. Proctor and K.-P. L. Vu, 284–300. Mahwah, NJ: Lawrence Erlbaum Associates.

Chua, H. F., Boland, J. E., and Nisbett, R. E. 2005. Cultural variation in eye movements during scene perception. http://www.pnas.org/cgi/content/abstract/102/35/12629 (Accessed October 7, 2005).

CNN. 2007. China's Web users could dominate. www.cnn.com/2006/TECH/internet/12/29/china.online.ap/index.html (Accessed January 24, 2007).

Columbia University. 1995. A survey of college and university WWW sites. http://www.ilt.columbia.edu/academic/classes/TU5020/projects/he/higher_ed.html (Accessed June 15, 2006).

Corfu, A., Lanranja, M., and Costa, C. 2003. Evaluation of tourism Web site effectiveness: Methodological issues and survey results. In *Human computer interaction: Theory and practice (Part I),* ed. J. A. Jacko and C. Stephanidis. Mahwah, NJ: Lawrence Erlbaum Associates, 753–757.

Culnan, Mary J. 1993. "How did they get my name?": An exploratory investigation of consumer attitudes toward secondary information use. *MIS Quarterly* 17:341–64.

Cyr, D., and Trevor-Smith, H. 2004. Localization of Web design: An empirical comparison of German, Japanese, and United States Web site characteristics. *Journal of the American Society for Information Science and Technology* 55:1199–1208.

De Mooij, M. 2000. The future is predictable for international marketers: Converging incomes lead to diverging consumer behavior. *International Marketing Review* 17:103–113.

De Mooij, M. 2003. Convergence and divergence in consumer behavior: Implications for global advertising. *International Journal of Advertising* 22:183–202.

De Mooij, M. 2004. Consumer behavior and culture: Consequences for global marketing and advertising. Thousand Oaks, CA: Sage.

Dedeke, A. 2000. A conceptual framework for developing quality measures for information systems. Paper presented at the 2000 International Conference on Information Quality, Boston, MA.

Degen, H., Lubin, K. L., Pedell, S., and Zheng, J. 2005. Travel planning on the Web: A cross-cultural case study. In *Usability and internationalization of information technology,* ed. N. Aykin. Mahwah, NJ: Lawrence Erlbaum Associates.

Delone, W. H., and McLean, E. R. 1992. Information systems success: The quest for the dependent variable. *Information System Research* 3(1):60–95.

Detlor, B., Sproule, S., and Gupta, C. 2003. Pre-purchase online information seeking: Search versus browse. *Journal of Electronic Commerce Research* 4(2):72–84.

Efendioglu, A. M., and Yip, V. F. 2004. Chinese culture and e-commerce: An exploratory study. *Interacting with Computers* 16:45–62.

EMarketer. 2008. Where is Generation X? http://www.emarketer.com/Article.aspx?id=1006699 (Accessed October 10, 2008).

Fang, X., and Salvendy, G. 2003. Customer-centered rules for design of e-commerce Web sites. *Communications of the ACM* 46:332–36.

Fitts, P. M. 1954. The informational capacity of the human motor system in controlling the amplitude of movements. *Journal of Experimental Psychology* 47:381–91.

Forsythe, S. M., and Shi, B. 2003. Consumer patronage and risk perceptions in Internet shopping. *Journal of Business Research* 56(11):867–75.

Franke, G. R., Huhmann, B. A., and Mothersbaugh, D. L. 2004. Information content and consumer readership of print ads: A comparison of search and experience products. *Journal of the Academy of Marketing Science* 32(1):20–31.

Frascara, J. 2000. Information design and cultural difference. *Information Design Journal* 9:119–27.

Gefen, D. 2000. E-commerce: The role of familiarity and trust. *Omega* 28:725–37.

Gehrke, D., and Turban, E. 1999. Determinants of successful Web site design: Relative importance and recommendations for effectiveness. Paper presented at the 1999 Proceedings of the Thirty-second Annual Hawaii International Conference on System Sciences, Maui, Hawaii.

Girard, T., Silverblatt, R., and Korgaonkar, P. 2002. Influence of product class on preference for shopping on the Internet. *Journal of Computer-Mediated Communication* 8(1):1–35. http://www.ascusc.org/jcmc/vol8/issue1/girard.html (Accessed October 12, 2004).

Gold, K. 2006. Do security measures boost conversion? http://www.practicalecommerce.com/articles/356-Do-Security-Measures-Boost-Conversion (Accessed September 20, 2008).

Graphics, Visualization, and Usability Center (GVU). 1998. GVU's 10th WWW User Survey. http://www.gvu.gatech.edu/user_surveys/survey-1998–10/ (Accessed February 3, 2005).

Greer, J. 2003. Evaluating the credibility of online information: A test of source and advertising influence. *Mass Communication and Society* 6(1):11–28.

Griffin, G. 1999. A typology of online positioning strategies among creative programs. http://www.ciadvertising.org/studies/student/99_fall/phd/griffin/onlinepaper/abstract.html (Accessed January 9, 2006).

Guo, Y., Proctor, R. W., and Salvendy G. Forthcoming. Content structure for information appliance: A survey study based on Chinese population.

Guo, Y., and Salvendy, G. 2007. Factor structure of content preparation for e-business Web sites: A survey results of 428 industrial employees in P. R. China. In *Proceedings of the 2007 Human-Computer Interaction (HCI) International (Part I),* ed. J. Jacko, 784–95. Heidelberg: Springer.

Guo, Y., and Salvendy, G. 2009. Factor structure of content preparation for e-business Web sites: Results of a survey of 428 industrial employees in P. R. China. *Behaviour & Information Technology* 28(1):73–86.

Hall, E. T. 1983. The dance of life: The other dimension of time. Yarmouth, ME: Intercultural.

Hedden, T., Park, D. C., Nisbett, R. E., Ji, L.-J., Jing, Q., and Jiao, S. 2002. Cultural variation in verbal versus spatial neuropsychological function across the life span. *Neuropsychology* 16:65–73.

Hilsenrath, J. 2002. The economy: Consumer satisfaction is rising—lower prices translate into sense of more value from products, services. *Wall Street Journal.* November 18.

Ho, J. 1997. Evaluating the World Wide Web: A global study of commercial sites. *Journal of Computer Mediated Communication* 3(1). http://jcmc.indiana.edu/vol3/issue1/ho.html (Accessed October 17, 2008).

Hofstede, G. H. 2001. *Culture's consequences: Comparing values, behaviors, institutions and organizations across nations.* 2nd ed. Thousand Oaks, CA: Sage.

Honold, P. 1999. Learning how to use a cellular phone: comparison between German and Chinese users. *Technical Communication* 46:196–205.

Hornbaek, K. 2006. Current practice in measuring usability: Challenges to usability studies and research. *International Journal of Human Computer Studies* 64(2):79–102.

Household income in the United States. n.d. In Wikipedia, the Free Encyclopedia. http://en.wikipedia.org/wiki/Household_income_in_the_United_States (Accessed In October 15, 2008).

Huang, K.-T., Lee, Y. W., and Wang, R. Y. 1999. *Quality information and knowledge.* Upper Saddle River, NJ: Prentice Hall PTR.

Hwang, W., Jung, H.-S., and Salvendy G. 2006. Internationalisation of e-commerce: A comparison of online shopping preferences among Korean, Turkish and US populations. *Behaviour & Information Technology* 25:3–18.

IBM. 2004. IBM Web design guidelines for e-commerce. http://www-306.ibm.com/ibm/easy/eou_ext.nsf/publish/611 (Accessed January 3, 2005).

Imation. 2001. Imation tames unruly Web color with Verifi technology. http://ir.imation.com/phoenix.zhtml?c=73967&p=irol-newsArticle&ID=567042&highlight (Accessed December 15, 2006).

International Organization for Standardization. 1991. ISO 9126. Software product evaluation—Quality characteristics and guidelines for their use.

International Organization for Standardization. 1998. ISO 9241-11. Ergonomic requirements for office work with visual display terminals (VDTs)—Part 11: Guidance on usability. http://www.idemployee.id.tue.nl/g.w.m.rauterberg/lecturenotes/ISO9241part11.pdf (Accessed July 28, 2009).

Internet World Stats. 2004. China Internet market brief. http://www.internetworldstats.com/asia/cn.htm (Accessed July 1, 2006).

iResearch. 2004. China online shopping research report (simple version). http://english.iresearch.com.cn/e_commerce/ (Accessed July 1, 2006).

Ives, B., Olson, M. H., and Baroudi, J. J. 1983. The measurement of user information satisfaction. *Communications of the ACM* 26(10):785–93.

Jackson, L. A., Eye, A. V., Barbatsis, G., Biocca, F., and Zhao, Y. 2004. The impact of internet use on the other side of the digital divide. *Communication of the ACM* 47(7):43–47.

Jackson, L. A., Eron, K. S., Gardner, P. D., and Schmitte, N. 2001. Gender and the Internet: Women communicating and men searching. Sex Roles 44:363–379.

Jarvenpaa, S. L., Tractinsky, N., and Saarinen, L. 1999. Consumer trust in an Internet store: A cross-cultural validation. *Journal of Computer-Mediated Communication* 5. http://jcmc.indiana.edu/vol5/issue2/jarvenpaa.html (Accessed February 3, 2005).

Jensen, A. R., and Whang, P. A. 1994. Speed of accessing arithmetic facts in long-term memory: A comparison of Chinese-American and Anglo-American children. *Contemporary Educational Psychology* 19:1–12.

Ji, L.-J., Peng, K., and Nisbett, R. E. 2000. Culture, control, and perception of relationships in the environment. *Journal of Personality and Social Psychology* 78:943–55.

Ji, L.-J., Zhang, Z., and Nisbett, R. E. 2004. Is it culture or is it language? Examination of language effects in cross-cultural research on categorization. *Journal of Personality and Social Psychology* 87:57–65.

Ji, M. F., and McNeal, J. U. 2001. How Chinese children's commercials differ from those of the United States: A content analysis. *Journal of Advertising* 30:79–92.

Ji, Y. G., Park, J. H., Lee, C., and Yun, M. H. 2006. A usability checklist for the usability evaluation of mobile phone user interface. *International Journal of Human-Computer Interaction* 20(3):207–31.

Johnson, K. T. 2008. Process, preference and performance: The role of ethnicity and socio-economics status in computer interface metaphor design. PhD diss., Virginia Polytechnic Institute and State University.

Jones, M. Y., Pentecost, R., and Requena, G. 2005. Memory for advertising and information content: Comparing the printed page to the computer screen. *Psychology and Marketing* 22(8):623–48.

Ju-Pak, K. 1999. Content dimensions of Web advertising: A cross-national comparison. *International Journal of Advertising* 18(2):207–30.

Kagan, J., Moss, H. A., and Sigel, I. E. 1963. *The psychological significance of styles of conceptualization.* Monograph of the Society for Research in Child Development 28, 73–112.

Kahn, B. K., Strong, D. M., and Wang, R. Y. 2002. Information quality benchmarks: product and service performance. *Communications of the ACM* 45(4):84–192.

Kaikkonen, A., Kekäläinen, A., Cankar, M., Kallio, T., and Kankainen, A. 2005. Usability testing of mobile applications: A comparison between laboratory and field testing. *Journal of Usability Studies* 1(1):4–17.

Karagiogoudi, S., Karatzas, E., and Papatheodorou, T. 2003. Web site quality evaluation: A case study on European cultural Web sites. In *Human computer interaction: Theory and practice (Part I)*, ed. J. A. Jacko and C. Stephanidis. Mahwah, NJ: Lawrence Erlbaum Associates, 783–787.

Katerattanakul, P., and Siau, K. 1999. Measuring information quality of Web sites: Development of an instrument. Paper presented at 20th International Conference on Information Systems, Charlotte, NC.

Klein, B. D. 2002. When do users detect information quality problems on the World Wide Web. Paper presented at the 2002 American Conference in Information Systems, Dallas, TX.

Klein, L. R. 1998. Evaluating the potential of interactive media through a new lens: Search versus experience goods. *Journal of Business Research* 41:195–203.

Knight, J. L., and Salvendy, G. 1992. Psychomotor work capabilities. In *Handbook of industrial engineering.* 2nd ed., ed. G. Salvendy. Hoboken, NY: Wiley, John and Sons.

Knight, S. A., and Burn, J. 2005. Developing a framework for assessing information quality on the World Wide Web. *Informing Science Journal* 8:160–72.

Kondratova, I., and Goldfarb, I. 2006. Cultural interface design: Global colors study. *Lecture Notes in Computer Science* 4278: 926–34.

Korgen, K., Odell, P., and Schumacher, P. 2001. Internet use among college students: Are there differences by race/ethnicity? *Electronic Journal of Sociology* 5(3). http: www.sociology.org/content/vol005.003/korgen.html (Accessed January 16, 2009).

Krauss, K. 2003. Testing an e-government Web site quality questionnaire: A pilot study. Paper presented at 5th Annual Conference on World Wide Web Applications, Durban, South Africa.

Kwon, O., Kim, C., and Lee, E. 2002. Impact of Web site information design factors on consumer ratings of Web-based auction sites. *Behaviour & Information Technology* 21(6):387–402.

Lazar, J., and Sears, A. 2006. Design of e-business Web sites. In *Handbook of human factors and ergonomics.* 3rd ed., ed. G. Salvendy. Hoboken, NJ: John Wiley and Sons, 1344–1363.

Leifer, A. 1972. Ethnic patterns in cognitive tasks. *Dissertation Abstracts International* 33:1270–71.

Lesser, G. S., Fifer, G., and Clark, D. H. 1965. *Mental abilities of children from different social class and cultural groups.* Monograph of the Society for Research in Child Development 30, part 4.

Lewis, J. R. 2006. Sample sizes for usability tests: Mostly math, not magic. *Interactions* 13(6):29–33.

Liao, H., Proctor, R. W., and Salvendy, G. 2008. Content preparation for cross-cultural e-commerce: A review and a model. *Behaviour & Information Technology* 27(1):43–61.

Liao, H., Proctor, R. W., and Salvendy, G. 2009. Chinese and US online consumers' preferences for content of e-commerce Web sites: A survey. *Theoretical Issues in Ergonomics Science,* 10:19–42.

Liao, H., Proctor, R. W., and Salvendy, G. In press. Content preparation for e-commerce involving Chinese and US online consumers. *International Journal of Human-Computer Interaction.*

Lightner, N. J. 2003. What users want in e-commerce design: Effects of age, education and income. *Ergonomics* 46(1–3):153–68.

Lin, C. 2001. Cultural values reflected in Chinese and American television advertising. *Journal of Advertising* 30:83–94.

Liu, Y., and Salvendy, G. 2009. Effects of measurement errors on psychometric measurements in ergonomics studies: Implications for correlations, ANOVA, linear regression, factor analysis, and linear discriminant analysis. *Ergonomics*, 52(5): 499–511.

Loiacono, E. T. 2000. WebQualTM: A Web site quality instrument. PhD diss., University of Georgia.

Lynn, R., Pagliari, C., and Chan, J. 1988. Intelligence in Hong Kong measured for Spearman's G and the visuospatial and verbal primaries. *Intelligence* 12:423–33.

Madden, C. S., Caballero, M. and Matsukubo, S. 1986. Analysis of information content in US and Japanese magazine advertising. *Journal of Advertising* 15:38–45.

Maes, P., Guttman, R. H., and Moukas, A. G. 1999. Agents that buy and sell. *Communications of the ACM* 42:81–91.

Marks, A., and Scherer, R. n.d. US savings rate falls to zero. http://articles.moneycentral.msn.com/Investing/Extra/USSavingsRateFallsToZero.aspx (Accessed October 10, 2008).

Masuda, T., and Nisbett, R. E. 2001. Attending holistically versus analytically: Comparing the context sensitivity of Japanese and Americans. *Journal of Personality and Social Psychology* 81:922–34.

McCort, D. J., and Malhotra, N. K. 1993. Culture and consumer behavior: Toward and understanding of cross-cultural consumer behavior in international marketing. *Journal of International Consumer Marketing* 6:91–127.

McKinney, V. Y. K., and Zahedi, F. M. 2002. The measurement of Web-customer satisfaction: An expectation and disconfirmation approach. *Information Systems Research* 13(3):296–315.

McKinsey (China) Global Institute. 2006. Putting China's capital to work: The value of financial system reform. http://www.mckinsey.com/mgi/publications/china_capital/index.asp (Accessed July 1, 2006).

Mello Jr., J. P. 2005. What are businesses doing with your personal data? http://www.ecommercetimes.com/story/45739.htm (Accessed September 26, 2008).

Miller, K. F., and Stigler, J. W. 1987. Counting in Chinese: Cultural variation in a basis cognitive skill. *Cognitive Development* 2:279–305.

Miniwatts Marketing Group. 2008. Internet World Stats. http://www.internetworldstats.com (Accessed November 7, 2008).

Miuar, I. T., Kim, C. C., Chang, C. M., and Okamoto, Y. 1988. Effects of language characteristics of children's cognitive representation of number: Cross-national comparisons. *Child Development* 59:1445–50.

Morris, M. W., and Peng, K. 1994. Culture and cause: American and Chinese attributions for social and physical events. *Journal of Personality and Social Psychology* 67:949–71.

Mossberger, K., Tolbert, C. J., and Gilbert, M. 2006. Race, place, and information technology. *Urban Affairs Review* 41(5):583–620.

Mowen, J. C., and Minor, M. 1998. *Consumer behavior.* 5th ed. Upper Saddle River, NJ: Prentice-Hall Inc.

Mueller, B. 1987. Reflections of culture: An analysis of Japanese and American advertising appeals. *Journal of Advertising Research* 27:51–59.

———. 1991. An analysis of information content in standardized vs specialized multinational advertisements. *Journal of International Studies* 22:23–39.

Naumann, F., and Rolker, C. 2000. Assessment methods for information quality criteria. Paper presented at the 2000 International Conference on Information Quality (IQ), Cambridge, MA.

Nielsen, J. 1992. Finding usability problems through heuristic evaluation. Paper presented at the 1992 ACM Conference on Human Factors in Computing Systems, Monterey, CA.

Nielsen, J., Snyder, C., Molich, R., and Farrell, S. 2001. *E-commerce user experience.* Fremont, CA: Nielsen Norman Group.

Nisbett, R. E. 2003. *The geography of thought: How Asians and Westerners think differently . . . and why.* New York: Free Press.

Nisbett, R. E., Peng, K., Choi, I., and Norenzayan, A. 2001. Culture and systems of thought: Holistic versus analytic cognition. *Psychological Review* 108:291–310.

Nitse, P. S., Parker, K. R., Krumwiede, D., and Ottaway, T. 2004. The impact of color in the e-commerce marketing of fashions: An exploratory study. *European Journal of Marketing* 38:898–915.

Noel-Levitz. 2005. Engaging the "Social Networking" generation: How to talk to today's college-bound juniors and seniors. https://www.noellevitz.com/NR/rdonlyres/425D56C3–9ACD-4-A90–9782-F70ED7AC3CF2/0/EExpectationsClassof2007.pdf (Accessed June 15, 2006).

Norman, D. A. 1988. *The psychology of everyday things.* New York: Basic Books.

Norton, S., and Norton Jr., W. 1988. An economic perspective on the information content of magazine advertisements. *International Journal of Advertising* 7(2):138–48.

Okazaki, S. 2004. Does culture matter? Identifying cross-national dimensions in Japanese mutinationals' product-based Web sites. *Electronic Markets* 14:58–69.

Okazaki, S., and Rivas, J. A. 2002. A content analysis of multinationals' Web communication strategies: Cross-cultural research framework and pre-testing. *Internet Research: Electronic Networking Applications and Policy* 12:380–90.

Onkvisit, S., and Shaw, J. 1999. Standardized international advertising: Some research issues and implications. *Journal of Advertising Research* 39(6):19–24.

Online shopping. n.d. In Wikipedia, the Free Encyclopedia. http http://en.wikipedia.org/wiki/Online_shopping (Accessed October 15, 2008).

Ono, H., and Zavodny, M. 2003. Race, Internet usage, and e-commerce. *Review of Black Political Economy* 30(3):7–22.

Pew Internet and American Life Project. 2005a. How women and men use the Internet. http://www.pewinternet.org/pdfs/PIP_Women_and_Men_online.pdf (Accessed October 10, 2008).

Pew Internet and American Life Project. 2005b. Reports: demographics. http://www.pewinternet.org/PPF/r/171/report_display.asp (Accessed October 10, 2008).

Pipino, L. L., Lee, Y. W., and Wang, R. Y. 2002. Data quality assessment. *Communications of the ACM* 45(4):211–18.

Plocher, T. A., and Zhao, C. 2002. Photo interview approach to understanding independent living needs of elderly Chinese: A case study. Paper presented at 5th Asia-Pacific Conference on Computer-Human Interface, Beijing, China.

Privacy Rights Clearinghouse. 2007. How many identity theft victims are there? What is the impact on victims? http://www.privacyrights.org/ar/idtheftsurveys.htm (Accessed October 3, 2008).

Proctor, R. W., Vu, K.-P. L., Najjar, L. J., Vaughan, M. W., and Salvendy, G. 2003. Content preparation and management for e-commerce Web sites. *Communications of the ACM* 46:289–99.

Proctor, R. W., Vu, K.-P., Salvendy, G., et al. 2002. Content preparation and management for Web design: Eliciting, structuring, searching, and displaying information. *International Journal of Human-Computer Interaction* 14(1):25–92.

Ranganathan, C., and Ganapathy, S. 2002. Key dimensions of business-to-consumer Web sites. *Information and Management* 39:457–65.

Ransom, K. 2008. Buy American as "economic patriotism"? http://www.cnn.com/2008/LIVING/wayoflife/09/12/aa.buy.american/ (Accessed October 2, 2008).

Ratchford, B. T., Talukdar, D., and Lee, M. 2001. A model of consumer choice of the internet as information source. *Journal of Electronic Commerce* 5:7–21.

Reibstein, D. J. 2002. What attracts customers to online stores, and what keeps them coming back? *Journal of the Academy of Marketing Science* 30(4):465–73.

Resnik, A., and Stern, B. L. 1977. An analysis of information content in television advertising. *Journal of Marketing* 41:50–53.

Rice, M. D., and Lu, Z. 1988. A content analysis of Chinese magazine advertisements. *Journal of Advertising* 17:43–48.

Roach, S. 2006. The US and China's savings problem. http://money.cnn.com/2006/03/03/news/international/chinasaving_fortune/ (Accessed October 13, 2008).

Rohn, J. 1998. Creating usable e-commerce sites. *StandardView* 6:110–115.

Ryu, Y. S. 2005. Development of usability questionnaires for electronic mobile products and decision making methods. PhD diss., Virginia Polytechnic Institute and State University.

Salvendy, G., and Seymour, W. D. 1973. *Prediction and development of industrial work performance.* Hoboken, NJ: John Wiley and Sons.

Sampei, M. A., Novo, N. F., Juliano, Y., Colugnati, F. A. B., and Sigulem, D. M. 2003. Anthropometry and body composition in ethnic Japanese and Caucasian adolescent girls: Considerations on ethnicity and menarche. *International Journal of Obesity* 27(9):1114–20.

Savoy, A., and Salvendy, G. 2007. Effectiveness of content preparation in information technology operations: Synopsis of a working paper. *Proceedings of the 2007 Human-Computer Interaction (HCI) International (Part I),* ed. J. Jacko, 624–31. Heidelberg: Springer

Savoy, A., and Salvendy, G. 2008. Foundations of content preparation for the Web. *Theoretical Issues in Ergonomics Science* 9(6):501–21.

Shackel, B. 1991. Usability-context, framework, definition, design and evaluation. In *Human factors for informatics usability*, ed. B. Shackel and S. J. Richardson. Cambridge: Cambridge University Press.

Shiller, R. 2006. The difference in saving rates between China and the US http://economistsview.typepad.com/economistsview/2006/08/differences_in_.html (Accessed October 13, 2008).

Smith-Jackson, T. L., Nussbaum, M. A., and Mooney, A. M. 2003. Accessible cell phone design: Development and application of a needs analysis framework. *Disability and Rehabilitation* 25(10):549–60.

Spooner, T., and Rainie, L. 2000. African-Americans and the Internet. http://www.pewinternet.org (Accessed November 10, 2008).

Sproull, L., and Kiesler, S. 1991. *Connections: New ways of working in the networked organization.* Cambridge, MA: MIT Press.

Statistical monthly report for the Chinese electronic and information industry July 2008. 2008. (In Chinese). http://www.hbh-info.gov.cn/xwdt/xwdt_4890.html (Accessed November 11, 2008).

Stigler, J. W., Lee, S. Y., and Stevenson, H. W. 1986. Digit memory in Chinese and English: Evidence for a temporally limited store. *Cognition* 23:1–20.

Strong, D. M., Lee, Y. W., and Wang, R. Y. 1997. 10 potholes in the road to information quality. *Computer* 30(8):38–46.

Szymanski, D. M., and Hise, R. T. 2000. E-satisfaction: An initial examination. *Journal of Retailing* 76(3):309–22.

Taylor, C. R., Miracle, G. E., and Wilson, R. D. 1997. The impact of information level on the effectiveness of US and Korean television commercials. *Journal of Advertising* 26:1–18.

Toya, E. S. 2008. American savings rate remains inadequate. February 26. http://blog.leonardwealthmanagement.com/lwm_blog/2008/02/american-saving.html. (Accessed December 5, 2008.)

Tullis, T. S., Catani, M., Chadwick-Dias, A., and Cianchette, C. 2005. Presentation of information. In *Handbook of human factors in Web design,* ed. R. W. Proctor and K.-P. L. Vu. Mahwah, NJ: Lawrence Erlbaum Associates, 107–133.

United Nations Conference on Trade and Development (UNCTAD). 2001. E-commerce and development report 2001. http://r0.unctad.org/ecommerce/docs/edr01_en.htm (Accessed July 1, 2006).

United Nations Conference on Trade and Development (UNCTAD). 2002. E-commerce and development report 2002. http://r0.unctad.org/ecommerce/ecommerce_en/edr02_en.htm (Accessed July 1, 2006).

Unsworth, S. J., Sears, C. R., and Pexman, P. M. 2005. Cultural influences on categorization processes. *Journal of Cross-Cultural Psychology* 36:662–88.

Van Duyne, D. K., Landay, J. A., and Hong, J. I. 2003. The design of sites: Patterns, principles, and processes for crafting a customer-centered Web experience. Boston: Addison-Wesley.

Vu, K.-P. L., and Proctor, R. W. 2006. Web site design and evaluation. In *Handbook of human factors and ergonomics.* 3rd ed., ed. G. Salvendy. Hoboken, NJ: John Wiley and Sons.

Walker, Q. R. 2002. Advertising and the internet. PhD diss., Northwestern University.

Wang, R. Y., and Strong, D. M. 1996. Beyond accuracy: What data quality means to data consumers. *Journal of Management Information Systems* 12(4):5–33.

Wang, Y., and Tang, T. 2001. An instrument for measuring customer satisfaction towards Web sites that market digital products and services. *Journal of Electronic Commerce Research* 2(3):1–28.

Westin, A. F. 1991. *Harris-Equifax consumer privacy survey 1991.* Atlanta: Equifax Inc.

Westin, A. F. 1998. *E-commerce and privacy: What Net users want.* Hackensack, NJ: Privacy & American Business.

World Bank, the. n.d. China quick facts. http://web.worldbank.org/WBSITE/EXTERNAL/COUNTRIES/EASTASIAPACIFICEXT/CHINAEXTN/0,,contentMDK:20680895~pagePK:1497618~piPK:217854~theSitePK:318950,00.html (Accessed October 15, 2008).

Zhang, D., and Adipat, B. 2005. Challenges, methodologies, and issues in the usability testing of mobile applications. *International Journal of Human-Computer Interaction* 18(3):293–308.

Zhang, L. 2005. Chinese credit card market may lead to losses. http://finance.sina.com.cn/money/bank/bank_card/20051029/10332077352.shtml (Accessed July 1, 2006).

Zhang, P., Von Dran, G., Blake, P., and Pipithsuksunt, P. 2001. Important design features in different Web site domains: An empirical study of user perceptions. *e-Service Journal* 1(1):77–91.

Zhang, P., von Dran, G., Small, R. V., and Barcellos, S. 1999. Web sites that satisfy users: A theoretical framework for Web user interface design and evaluation. Paper presented at 32nd Hawaii International Conference on System Sciences, Maui, Hawaii.

Zhang, Y., Chen, J. Q., and Wen, K.-W. 2002. Characteristics of Internet users and their privacy concerns: A comparative study between China and the United States. *Journal of Internet Commerce* 1:1–16.

Index

G

H

I

J

K

L

M